AF732300

This book is to be returned on or before
the last date stamped below.

Renewals
BUCKINGHAM (F)
01280 828267

LONG LOAN

6/6/08

10 OCT 2008

16 FEB 2009

10 JUN 2009

07 OCT 2010

LIBREX

LATE RETURN WILL INCUR
PENALTY \ FINE

Naoyuki Osaka, Ingo Rentschler, Irving Biederman (Eds.)

Object Recognition, Attention, and Action

Naoyuki Osaka, Ingo Rentschler,
Irving Biederman (Eds.)

Object Recognition, Attention, and Action

With 77 Figures

Springer

Naoyuki Osaka, Ph.D.
Professor, Department of Psychology, Graduate School of Letters, Kyoto University
Yoshida-Honmachi, Sakyo-ku, Kyoto 606-8501, Japan

Ingo Rentschler, Ph.D.
Professor, Institute of Medical Psychology, University of Munich
Goethestr. 31, D-80336 Munich, Germany

Irving Biederman, Ph.D.
Professor, Image Understanding Laboratory, University of Southern California
Hedco Neurosciences Building, Rm. 316, MC 2520, 3641 Watt Way, Los Angeles, CA 90089-2520, USA

LIBRARY
08119056
006.
37
OSA

ISBN 978-4-431-73018-7 Springer Tokyo Berlin Heidelberg New York

Library of Congress Control Number: 2007931334

Printed on acid-free paper

© Springer 2007
Printed in Japan
This work is subject to copyright. All rights are reserved, whether the whole or part of the material is concerned, specifically the rights of translation, reprinting, reuse of illustrations, recitation, broadcasting, reproduction on microfilms or in other ways, and storage in data banks.
The use of registered names, trademarks, etc. in this publication does not imply, even in the absence of a specific statement, that such names are exempt from the relevant protective laws and regulations and therefore free for general use.

Springer is a part of Springer Science+Business Media
springer.com

Typesetting: SNP Best-set Typesetter Ltd., Hong Kong
Printing and binding: Shinano Inc., Japan

Preface

Visual object recognition has been for years one of the most intensively studied subjects in cognitive science. It is only recently, however, that scientists have been able to investigate the neuronal processes possibly underlying this basic cognitive ability. Recent progress in cognitive/computational neuroscience and visual psychophysics has allowed further understanding of the neuronal and behavioral correlates associated with the different forms of object recognition.

This volume provides a comprehensive view of the neuronal and behavioral bases of object recognition, taking as its thesis that object recognition involves both active attention and coordinated action to adapt to the world. To fully understand human object recognition, therefore, we are required to examine its concept from the multidisciplinary point of view involving psychophysical research as well as cognitive and computational neuroscience of attentional mechanisms and action.

The collection of articles in this volume is produced on the basis of talks and in-depth discussions at the International Workshop on Object Recognition, Attention, and Action, organized by Naoyuki Osaka (Kyoto University, Japan) and Ingo Rentschler (University of Munich, Germany), and held at Kyoto University in 2004.

Leading researchers on object recognition believe that a firmer understanding of this topic is now within our reach because of new evidence from cognitive neuroscience, cognitive science, and neuropsychology. Accordingly, the neuronal system supporting object recognition seems to be in attentional networks connecting the visual brain with temporo-parietal cortex and even the prefrontal cortex. Furthermore, the coordination across various brain areas probably serves the purpose of binding purposeful action to the recognition task at hand. The present volume is to provide a forum for systematic comparison of present models and theories of object recognition in the brain. Thus, it aims at encouraging communication between students and researchers from different subdisciplines of cognitive science by focusing on explicit, detailed comparisons of current major approaches to object recognition theory and modeling. The domains in which the present contributors have examined the role of the neuronal basis of object recognition include higher brain mechanisms, attention, perception, working memory, binding in the cerebral cortex, neural networks, and voluntary action.

As the book covers a wide range of different theoretical perspectives and interdisciplinary views, it will be of interest also to researchers and students in cognitive science/psychology, cognitive neuroscience, neuropsychology, neurobiology, artificial intelligence, and philosophy of the mind.

Acknowledgments

The International Workshop on Object Recognition, Attention, and Action that inspired this book was organized by the International Steering Committee of the workshop, directed by Naoyuki Osaka (Japan) and Ingo Rentschler (Germany). The workshop was held August 3–6, 2004, at the Centennial Conference Hall, Kyoto University. The International Scientific Committee for the workshop comprised Naoyuki Osaka (Chair, Kyoto University), Ingo Rentschler (University of Munich), and Irving Biederman (University of Southern California).

We are grateful to our session chairs Keiji Tanaka, Irving Biederman, Ingo Rentschler, Heinrich Bülthoff, and Jules Davidoff and to the following contributors: Hiroshi Ando, Hiroshi Ashida, Irving Biederman, Heinrich Bülthoff, Jules Davidoff, Gustavo Deco, Jiro Gyoba, Martin Jüttner, Nobuyuki Hirose, Ken Kihara, Mutsutaka Kobayakawa, Zili Liu, Yoshihiro Miyake, Yoshitaka Ohigashi, Naoyuki Osaka, Ingo Rentschler, Jun Saiki, Sophie Schwartz, Hiroyuki Sogo, Hiroshi Shibata, Hans Strasburger, Bosco Tjan, Keiji Tanaka, Shigeki Tanaka, and Patrik Vuilleumier.

We also express our thanks to the conference staff including Nobuyuki Hirose, Takashi Ikeda, Mizuki Kaneda, Ken Kihara, Daisuke Matsuyoshi, Rie Nishimura, Yuki Otsuka, Hiroyuki Sogo, and Hiroyuki Tsubomi from the Osaka Lab, Kyoto University.

We are deeply indebted to the editorial staff members of Springer Japan for their enduring support and for their assistance in assembling the manuscript.

Finally, many thanks go to the Japan Society for the Promotion of Sciences (JSPS), the Kyoto University 21st Century COE Program (D-2, Kyoto University) from MEXT Japan, the Kyoto University Foundation, the Inoue Science Foundation, and the Kayamori Foundation for their financial support.

Contents

Part II: Attention

Part III: Action

Contributors

Ashida, Hiroshi
Department of Psychology, Graduate School of Letters
Kyoto University, Japan

Biederman, Irving
Department of Psychology and Neuroscience Program
University of Southern California, USA

Brettel, Hans
CNRS UMR
École Nationale Supérieure des Télécommunications, France

Bülthoff, Heinrich
Cognitive and Computational Psychophysics
Max Planck Institute for Biological Cybernetics, Germany

Davidoff, Jules
Psychology Department
Goldsmiths University of London, UK

Deco, Gustavo
Department of Technology Computational Neuroscience
Universitat Pompeu Fabra, Spain

Gschwind, Markus
Institute of Medical Psychology
University of Munich, Germany

Jüttner, Martin
School of Life and Health Sciences – Psychology
Aston University, UK

Miyake, Yoshihiro
Department of Computational Intelligence and Systems Science
Tokyo Institute of Technology, Japan

Onishi, Yohei
Department of Computational Intelligence and Systems Science
Tokyo Institute of Technology, Japan

Osaka, Naoyuki
Department of Psychology, Graduate School of Letters
Kyoto University, Japan

Plomp, Gijs
Laboratory for Perceptual Dynamics
BSI RIKEN, Japan

Pöppel, Ernst
Institute of Medical Psychology
University of Munich, Germany

Rentschler, Ingo
Institute of Medical Psychology
University of Munich, Germany

Saiki, Jun
Graduate School of Human and Environmental Studies
Kyoto University, Japan

Schwartz, Sophie
Department of Neurosciences
University of Geneva, Switzerland

Sogo, Hiroyuki
Department of Psychology, Graduate School of Letters
Kyoto University, Japan

Stetter, Martin
Information & Communications
Siemens AG, Germany

Strasburger, Hans
Department of Medical Psychology
University of Göttingen, Germany

Szabo, Miruna
Department of Computer Science
Technical University of Munich, Germany

Tanaka, Shigeki
Department of Psychology
Jin-Ai University, Japan

Thoma, Volker
School of Psychology
University of East London, UK

van Leeuwen, Cees
Laboratory for Perceptual Dynamics
BSI RIKEN, Japan

Vuilleumier, Patrik
Department of Neurosciences
University of Geneva, Switzerland

Wallraven, Christian
Cognitive and Computational Psychophysics
Max Planck Institute for Biological Cybernetics, Germany

An Editorial Overview

Ingo Rentschler, Naoyuki Osaka, and Irving Biederman

1 Introduction

To paraphrase a famous statement by Isaac Newton (Hawking 2002), we stand on the shoulders of giants when we seek insights into how humans recognize objects within their world. Aristotle showed that objects are assigned to categories according to attributes they have in common with other occurrences (Russell 1961). Immanuel Kant contended that the objects of our intuition (German *Anschauung*) are not representations of things as they are in themselves but appearances shaped by relations to things unknown to our sensibility. Synthetic judgments are needed to bind these appearances together, but these processes do not entail cognition per se. According to Kant, the compounds become integrated and understood by their assignment to the categories of pure reason (Zöller 2004). Arthur Schopenhauer (1859), however, was willing to accept only causality as a category of understanding. To Schopenhauer, causality was conditional for any act of cognition.

Aristotle's view of object categorization remained unchallenged until Ludwig Wittgenstein asked how we are able to identify something as an instance of the category of "games." The philosopher wondered what a game of darts, for instance, might have in common with the game of soccer. His concept of "family resemblance" (German *Familienähnlichkeit*) replaced the Aristotelian idea of a certain set of attributes being common to all occurrences of a particular class. Accordingly, attributes are distributed across the members of a family, or category, in a probabilistic fashion. Thus, games, tables, and trees were natural families for Wittgenstein, each constituted by a crisscross network of overlapping resemblances (Glock 1996).

Wittgenstein's concept of categorization was seminal for the philosophy of science. Indeed, Thomas Kuhn proposed that normal science does not work according to certain objective rules. Instead, it rests on the ability of scientific communities to relate problems to model solutions or paradigms, i.e., class descriptions in the sense of technical pattern recognition. Thus, it is conceivable that scientific revolutions such as the transition from Newtonian to quantum mechanics are brought about by changes in paradigms. Familiar demonstrations of the ambiguities of visual gestalt are suggestive for characterizing such changes. Indeed, "what were ducks in the scientist's world before the revolution are rabbits afterwards." (Kuhn 1970, p.111).

Transitions of paradigms happen in cognitive science in much the same way as in any other domain of science. However, as an interdisciplinary science, it is plagued by changes that rarely occur in synchrony across its subdisciplines, thus causing disparities among these fields. The present collection of chapters entitled *Object Recognition, Attention, and Action* aims, therefore, at drawing attention to a number of developments that appear to have happened independently in these distinct areas of cognitive science. However, it might become clear that their consequences are never restricted to just one of these topics, thus indicating the existence of a common framework for their integration.

Concerning the relationships of object recognition and action, it has been suggested that these types of functions rely on separate processing streams in the brain such as the ventral ("what") and the dorsal ("where") pathways in the monkey (Ungerleider and Mishkin 1982). Milner and Goodale (1995) confirmed this concept for humans but were led to emphasize the role of the dorsal ("how to") stream in visually guided behavior. They also conjectured that there exist areas in the parietal cortex where information from both cortical streams as well as other sensory modalities is integrated for the formation of abstract spatial representations as are needed, for instance, in understanding maps (Milner and Goodale 1995, Sec. 4.5).

More recently, evidence has been accumulated that the posterior parietal cortex (PPC) forms multiple spatial maps of the world. In the monkey, PPC constructs multiple space representations related to specific classes of action (Rizzolatti and Arbib 1998; Matelli and Luppino 2001). According to a model by Fagg and Arbib (1998), one such representation provides visual descriptions of three-dimensional objects, thus "proposing" to one area of the premotor cortex several possibilities of grasping. The most appropriate grip is selected based on the current position of the hand (so that it is a hand coordinate space rather than a viewer-centered coordinate space) and contextual information, and this information is sent to another area of the premotor cortex for motor execution. When humans select between actions with regard to context, the temporal cortex and its prefrontal projections are involved as pathways (Fig. 1) (Passingham and Toni 2001).

The importance of context, i.e., information gathered at some other place or time (Albright 1995), for visual recognition is widely acknowledged. Nevertheless, surprisingly little is known about the neuronal mechanisms that mediate contextual effects and scene analysis (Bar 2004). Traditionally, context has been regarded as an independent source of information facilitating the interpretation of visual input information via association (Biederman et al. 1974; Massaro 1979). A different view prevails in machine vision, where contextual information has been used to build robust pre-processing schemes allowing for reliable extraction of features for object recognition and scene analysis (Clowes 1971; Freuder 1986; Caelli and Bischof 1997). Recurrent coupling between "world knowledge" and

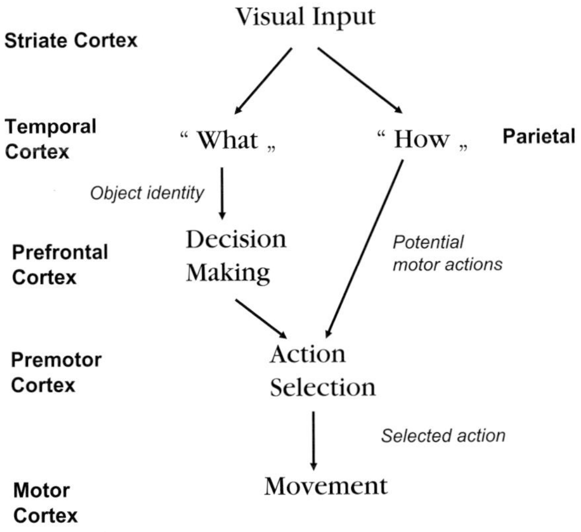

FIG. 1. Two visual streams connecting action and movement (Adapted with permission from Passingham and Toni 2001)

various processing stages is crucial for the latter approach, and there is reason to believe it also plays a key role in visual recognition (Lee et al. 1998; Bullier 2001; Briscoe 2000; Jüttner et al. 2004).

Attention was originally seen as a means of selecting stimulus dimensions or attributes (Allport 1980). This view was complemented by the concept of space-based attention functioning as a "mental spotlight," which focuses on one region of the visual field at a time (Eriksen and Hoffman 1973). More recently, there is evidence for biased competition (see the chapter by G. Deco, this volume), where at some point between input and response, input objects compete for representation, analysis, or control. Competition is biased in part by bottom-up mechanisms providing spatial and temporal segmentation of objects in a scene and in part by top-down mechanisms that select objects relevant for the current behavior. Accordingly, attention is not a high-speed device scanning each item in the visual field but is an emergent property of slow, competitive interactions working at various levels of brain function (Desimone and Duncan 1995). Because attention was believed to have a capacity-limited nature, it needed to be allocated in accordance with the executive function of working memory (Osaka et al. 2007).

Another disparity between subdisciplines of cognitive science concerns the role of learning in visual recognition. Psychophysicists tend to believe that recognition is mediated by feature extraction through fixed and invariant neural mechanisms. Accordingly, nothing is learned with respect to these mechanisms beyond, perhaps, infancy insofar as critical periods may exist for establishing experience-dependent neural connectivity. Others, concerned with the emergence of visual expertise, are inclined to assume that adult learning plays a major role in recognition. Nevertheless, recognition performance is assessed by both communities using paradigms of stimulus discrimination. Similarly, neural net models assume that class members share certain features, or feature vectors, and that classes can be separated by discriminative functions. Such concepts based on Aristotelian categorization are no longer useful when pattern complexity increases or patterns are embedded in scenes. For these reasons, artificial intelligence developed syntactic pattern recognition, where the similarity of patterns and class paradigms is measured using structural similarities as in graph matching (Grimson 1990; Bischof and Caelli 1997; Bunke 2000).

Concerning the relationships of object recognition, attention, and action, it has long been known that defects of afferent visual pathways, from retina to cortex, are instantly noticed by the patient who becomes aware of his or her loss of *shape-based recognition*. That is, the patient becomes aware of the inability to recognize people – or even to know that they are people – or objects. Bilateral occipital infarction, by contrast, may result in a loss of object recognition combined with a complete loss of visual imagery, visual memory as well as visual dreaming. Yet patients may remain unaware of these deficiencies. Similarly, following focal destruction of other cortical areas, submodalities of visual recognition may become lost, with the patient being unable to clearly communicate what happened. Such clinical observations suggest that the neuronal activity in different cortical areas is related to different submodalities of object recognition with the possibility of attention being coupled to visual processing (Baumgartner 1990; Grüsser and Landis 1991). Objectively, unilateral removal of the anterior temporal lobe does not result in any noticeable deficit in object recognition (Biederman et al. 1997). Thus, it may be that in humans, the representations mediating visual recognition are fully computed in areas posterior to the anterior temporal lobes, such as the lateral occipital complex (Malach et al. 1995). The more anterior areas of the temporal cortex in humans may be a repository of episodic memories in which perceptual representations are bound to particular perspectives and contexts. Visual imagery can be regarded as "playing back" these representations onto the "screens" of retinotopic areas. Taken together, these observations are indicative of the handshaking that can occur between perceptual representations and episodic memories.

When objects are compared sequentially, as in category learning, object representations must be stored in working memory. This is also the case when the goals and intentions of one task are maintained while performing another task. Such types of information updating and retention of stimulus-related information can be attributed to the function of the frontal lobes of the brain (Fletcher and Henson 2001). Moreover, *any* form of motor activity inevitably changes activation patterns in the sensory cortex. This results in novel patterns of sensation, imagery, and emotional activation, which are communicated to the frontal brain. Consequently, knowledge about the world and action towards it can be intimately inter-related in the "perception–action" cycle (Fuster 2001, 2003), although "couch potato" states, as when watching television with minimal action, may constitute the greater portion of human (vs. infra-human) information assimilation. Thus, the extent to which the perception–action cycle is fundamental to human cognition and the extent to which it can provide a unifying theme in the cognitive neuroscience of object recognition, attention, and action remain to be determined.

2 This Collection of Chapters

The present volume contains 15 chapters written by separate researchers or research groups, which focus on object recognition, attention, and action, respectively. Most of the chapters are devoted to human performance and its likely neural correlates, but there are also three reports on computational concepts (Jüttner; Wallraven and Bülthoff; Deco, Stetter, and Szabo) that help to elucidate characteristics of visual processing underlying human performance in object recognition.

2.1 Pattern and 3D Object Recognition

Plomp and van Leeuwen discuss the problem of perceptual occlusion that results from the projection of a three-dimensional (3D) world onto a two-dimensional (2D) surface, the retina. This projection annihilates a large amount of information that is important for object recognition. Yet this loss of information can be compensated for, at least partially, by varying the geometry of projection through action and using information gathered at some other time and place. Perceptual completion is, therefore, an active process that is critically dependent on contextual knowledge and intentional behavior.

By integrating behavioral and brain-imaging data, Schwartz addresses the question of how the brain selects information that enables visual recognition within a given context. Her results demonstrate that both perceptual learning and selective attention may enhance the processing of object information and reduce that of non-object information. Contrary to the traditional view of primary sensory cortices as hard wired devices, such processes are observed in the early visual cortex. They seem to originate from local interactions of neural mechanisms as well as top-down influences reflecting behavioral strategies.

Strasburger and Rentschler report that the inferiority of visual pattern recognition in indirect view is only partially explained by variations in spatial resolution across the visual field. Assuming a lack of feature integration or structural encoding on indirect view is also insufficient. The authors' recent experiments on pattern categorization on direct and indirect view suggest that objects are represented in the brain at several levels from the sensory to the conceptual, with spatial attention operating at an earlier level and object selective attention at a later level.

Jüttner employed part-based strategies from image understanding by computer to resolve the problem of structural pattern recognition in human vision. This approach goes beyond the application of standard neural nets or decision trees insofar as object attributes are not linked to patterns as wholes but to labeled pattern parts. The task of category learning is then to estimate what relations between pattern parts and their attributes best fit human recognition performance. This research strategy has provided quantitative details about the nature of structural representations.

Biederman contrasts feature hierarchies and structural descriptions as accounts of object (but not face) representations. The former types of representation are pixel-based in the sense that even though the pixels are mapped onto higher-level features, such as Gabor kernels or vertices, they can be "played back" to recover the pixel values. The latter may encode viewpoint-invariant geometric primitives derived from configurations of image edges (orientation and depth discontinuities), and their relations. A major argument in favor of structural descriptions is that they can mediate object recognition that is invariant to 3D rotations, lighting changes, contrast reversal, and partial occlusion. Returning to Wittgenstein's consideration of the nature of concepts, structural descriptions (SDs) offer an easy solution of how object categories that bear little visual resemblance to each other – such as chairs and lamps – might, nonetheless, be understood by a child. We can have multiple SDs per object category so that family resemblance need only be computed to the nearest SD, rather than all instances of that class. Recent evidence from neurophysiology and psychophysics supports the view that structural representations play an important role in visual object recognition.

Vuilleumier considers face recognition with the emphasis on identity and expression. Using brain imaging, he demonstrated that representations in the main brain regions associated with face recognition are neither view-invariant nor restricted to encode identity. Even for famous faces, invariant recognition involves semantic information from other brain structures, rather than "view-independent representations" per se. Similarly, there are emotional effects in cortical face regions that are generated by amygdala feedback. These findings show *pars pro toto*: higher brain function is the result of large-scale dynamic interactions of a number of neuronal populations.

2.2 Object Recognition and Attention

Thoma and Davidoff discuss and further test the performance of a hybrid model of visual object recognition and attention combining aspects of "view-based" recognition with the use of structural descriptions. Accordingly, object representations differ depending on whether objects are attended or not. "Holistic" (pixel-based) representations are formed with and without attention allowing for rapid recognition under limited invariance conditions. "Analytical" (structural) representations are built for attended objects only, permitting a larger extent of invariance to changes in viewpoint or shape distortion at the expense of processing time.

Sogo and Osaka argue that the influence of saccadic compression and illusory perception of an object location during the perisaccadic period on shape perception would show that there is an interaction between the "what" and "where" pathways. Perception of natural scenes, Glass pattern, and a Kanizsa subjective figure are affected by saccadic compression, while perception of a single object and an array of elements perceptually grouped in a single object are unaffected. Accordingly, saccadic compression originates in the "where" pathway, and thus seems to affect perception of object shapes through a feedforward–feedback loop between earlier and higher visual areas.

Saiki addresses the binding problem using multiple object permanence tracking designed to evaluate visual working memory. They tested whether prestored combinations (natural objects) or constant correspondences of shape and color facilitate memory for binding. Neither prior knowledge nor constant mapping had significant effects on accuracy in task performance, suggesting that limitation in binding memory is not an artifact of arbitrary feature combinations. With natural objects, people are sensitive to changes in color–shape combination, while shape, color, and location independently affect performance during observation of geometric figures, suggesting possible structural differences in memory representations.

Deco, Stetter, and Szabo address the problem of how representations held in different cortical areas might be integrated to form a coherent stream of perception, cognition, and action. They introduce the principle of biased competition and cooperation (BCC), allowing the modeling of attentional filtering, where competition and cooperation occur within a single model brain area. The operation of BCC across two different brain areas provides a model for visual category learning. Deco and co-authors further show how Hebbian synaptic plasticity can induce increased performance in a multi-area system.

2.3 Object Recognition and Action

Wallraven and Bülthoff contend that the extraction of structural view-invariant primitives may be too expensive computationally for 3D object recognition. Motivated by their own research on how observers recognize individuals from sequences of faces obtained by 3D laser-scans of heads, they explore the performance of machine recognition systems that combine an efficient pixel-based approach with a feature-tracking across time series of object views. The essence of such strategies, both in biological and machine vision, is the coupling of data from sensation and action, thus implementing perception–action cycles.

Gschwind, Brettel, and Rentschler show that structure-based visual categorization may be relatively easy for 3D objects composed of regularly shaped parts. However, replacing such parts by spherical parts renders categorization much more difficult. Indeed, spherical parts have ill-defined axes, thus lacking important information about macro-geometric object structure. Yet observers can resolve the resulting ambiguities concerning 3D structure using contextual information from prior active haptic exploration. The question of whether structural representations are used for visual object recognition cannot be decided, therefore, by analyzing image information only.

Ashida argues the influence of visual motion on perceptual and visuomotor performances during target localization. Visual illusions due to motion-related positional shifts that suggest visual extrapolation for a moving target were tested. Findings that positional shift is more prominent in visuomotor reaching tasks rather than in perceptual judgments support the theory of separate visual pathways for perception and action.

Tanaka addresses action imitation because imitation plays a critical role in human cognition. During action imitation, target actions are recognized visually and translated into one's own body representation, and finally imitated actions are performed using our own body. Mental representation of actions seems to play an important role in recognizing others' actions. A target action presented by another person's body is an object that we recognize using our own body representation. Findings from imaging studies showed different parietal contributions to finger action imitation and hand/arm action imitation.

Miyake, Onishi, and Pöppel introduce synchronization tapping to test the anticipatory-timing mechanism under single and dual-task. Findings indicate that tapping performance was affected in dual-task under an inter-stimulus interval of 1.5 to 3.5 s due to maintenance rehearsal involving the phonological loop of verbal working memory.

This collection of chapters will draw the reader's attention to a number of recent developments occurring in these specific areas of object recognition, attention, and action in cognitive neuroscience, cognitive psychology, and cognitive science.

References

Albright TD (1995) "My most true mind thus makes mine eye untrue". Trends in Neuroscience 18:331–333

Allport DA (1980) Attention and performance. In: Claxton G, ed. Cognitive Psychology: New Directions. Routledge & Kegan Paul, London, pp 112–153

Bar M (2004) Visual objects in context. Nature Reviews Neuroscience 5:617–629

Baumgartner G (1990) Where do visual signals become a perception? Pontificiae Academiae Scientiarum Scripta Varia 78:99–118

Biederman I, Gerhardstein PC, Cooper EE, Nelson CA (1997) High-level object recognition without an anterior inferior temporal cortex. Neuropsychologia 35:271–287

Biederman I, Rabinowitz J, Glass AL, Stacy, EW Jr (1974) On the information extracted from a glance at a scene. Journal of Experimental Psychology 103:597–600

Bischof WF, Caelli T (1997) Visual learning of patterns and objects. IEEE Transactions on Systems, Man, and Cybernetics – Part B: Cybernetics, 27:907–917

Briscoe G (2000) Vision as temporal trace. Spatial Vision 13:215–230

Bullier J (2001) Integrated model of visual processing. Brain Research Reviews 36:96–107

Bunke H (2000) Graph matching for visual object recognition. Spatial Vision 13:335–340

Caelli T, Bischof WF (1997) Machine Learning and Image Interpretation. Plenum Press, New York

Clowes MB (1971) On seeing things. Artificial Intelligence 2:79–116

Desimone R, Duncan J (1995). Neural mechanisms of selective visual attention. Annual Reviews Neuroscience 18:193–222

Eriksen CW, Hoffman JE (1973) The extent of processing of noise elements during selective encoding from visual displays. Perception & Psychophysics 14:155–160

Fagg AH, Arbib MA (1998) Modelling parietal-premotor interactions in primate control of grasping. Neural Networks 11:1277–1303

Fletcher PC, Henson RNA (2001) Frontal lobes and human memory. Brain 124:849–881

Freuder EC (1986) Knowledge mediated perception. In: Nusbaum HC, Schwab EC, eds. Pattern Recognition by Humans and Machines: Visual Perception. Academic Press, Orlando FL, pp 219–236

Fuster JM (2001) The prefrontal cortex – an update: time is of the essence. Neuron 30:319–333

Fuster JM (2003) Cortex and Mind. Oxford University Press, Oxford

Glock H-J (1996) A Wittgenstein Dictionary. Blackwell, Oxford

Grimson WEL (1990) Object Recognition by Computer: The Role of Geometric Constraints. MIT Press, Cambridge MA

Grüsser O-J, Landis T (1991) Visual Agnosias and Other Disturbances of Visual Perception and Cognition. Vol 12, Cronley-Dillon, J., ed. Vision and Visual Dysfunction. MacMillan, Houndmills

Hawking S, ed (2002) On the Shoulders of Giants. The Great Works of Physics and Astronomy. Running Press, Philadelphia

Jüttner M, Langguth B, Rentschler I (2004) The impact of context on pattern category learning and representation. Visual Cognition 11:921–945

Kuhn TS (1970) The Structure of Scientific Revolutions, 2nd edn. The University of Chicago Press, Chicago

Lee TS, Mumford D, Romero R, Lamme VAF (1998) The role of primary visual cortex in higher level vision. Vision Research 38:2429–2454

Malach R, Reppas JB, Benson RR, Kwong KK, Jlang H, Kennedy WA, Ledden PJ, Brady TJ, Rosen BR, Tootell RBH (1995) Object-related activity revealed by functional magnetic resonance imaging in human occipital cortex Proceedings of the National Academy of Science USA 92:8135–8139

Massaro DW (1979) Letter information and orthographic context in word perception. Journal of Experimental Psychology: Human Perception and Performance 5: 595–609

Matelli M, Luppino G (2001) Parietofrontal circuits for action and space perception in the macaque monkey. NeuroImage 14:S27–S32

Milner AD, Goodale MA (1995) The Visual Brain in Action. Oxford University Press, Oxford

Osaka N, Logie R, D'Esposito M, eds (2007) The Cognitive Neuroscience of Working Memory. Oxford University Press, Oxford

Passingham RE, Toni I (2001) Contrasting the dorsal and ventral visual systems: guidance of movement versus decision making. NeuroImage 14:S125–S131

Rizzolatti G, Arbib MA (1998) Language within our grasp. Trends in Neuroscience 21, 188–194

Russell B (1961) History of Western Philosophy, 2nd edn. George Allen & Unwin, London

Schopenhauer A (1859) Die Welt als Wille und Vorstellung. 3rd ed, vol 1, suppl. F A Brockhaus, Leipzig

Ungerleider LG, Mishkin A (1982) Two cortical systems. In: Analysis of Visual Behavior. Ingle DJ, Goodale MA, Mansfield RJW, eds. MIT Press, Cambridge MA, pp 549–586

Zöller G, ed (2004) Kant. Prolegomena to Any Future Metaphysics. Oxford University Press, Oxford

Part I
Object Recognition

1
Occlusion Awaits Disclosure

Gijs Plomp[1,2] and Cees van Leeuwen[1,2]

1 Perceptual Completion

At any moment in time, the world presents us with visual information that is inherently incomplete. This incompleteness arises to a large extent from the projection of a three-dimensional (3D) world onto a two-dimensional (2D) surface, the retina. The projected image fails to reveal, among other things, the way objects and surfaces extend behind others so that they appear partly occluded. From the observer's point of view anything whatsoever could be hidden behind an occluding object, including parts of the object itself; there is no principled way to derive what is there. Nonetheless we usually perceive occluded parts as having a determinate structure.

Occlusion is not the only problem resulting from a 2D projection of the 3D world. Another such problem is the size-distance invariance relation; when two objects project onto roughly the same area of the retina, this could mean that they are similar in size, but also that the larger one is further away. In principle, there is a continuity of possible solutions that meet the proportionality of distance and size. Yet, our visual system normally provides us with a definite, single, preferred solution.

The size-distance and occlusion problems share, it seems, an important characteristic: intrinsic uncertainties are quickly and quietly resolved by our visual system. It has been proposed that the visual system is selectively tuned to properties such as binocular disparity, relative size, familiar size and shading patterns that offer cues for resolving the size-depth invariance. For completing an occluded image, contour extrapolations, symmetry, similarity, proximity, and good volume continuation or complete mergeability (Tse 1999a, b) could be proposed as potentially relevant static cues.

[1]Laboratory for Perceptual Dynamics, BSI RIKEN, 2-1 Hirosawa, Wako-shi, Saitama 351-0198, Japan
[2]Business School, Sunderland University, St. Peter's Way, Sunderland SR6 0DD, UK

Importantly, additional information about the 3D world is provided by the way the environment changes with the continuous movement of our eyes, head, and body (Gibson 1972). We are able to detect and anticipate the dynamic properties of our environment by interacting with it. Size-distance invariance is broken when we move towards an object, quickly revealing its proportions. Even if we do not actually move, we have an expectation of what will happen when we do. We may consider the possibility that such anticipation has a special role in how our visual system decides perceived object size for us. Likewise, completion processes may be understood as anticipation of what happens when a momentary occluded object is disclosed, through parallax motion or relative displacement of the occluded and occluding objects, for instance.

Dynamic interaction of self and environmental, however, cannot entirely undo the inherent uncertainty in perception. Retinal expansion of an approaching object could, in principle, still be understood as an effect of that object growing in size while remaining equidistant. Disclosure of an occluded part can be considered, alternatively, as an un-occluded object growing a new part. These alternative solutions are very unlikely but a non question-begging account of perception will have to explain why the visual system discards them. Such a theory is still wanting. Advances can only be made on the basis of an appropriate description of the problem.

In the present chapter, we will elaborate the description of perceptual completion. We will do so in reference to the proposal that anticipated disclosure plays a role in how the occluded object is perceived. Such a description requires that completion is an active process, influenced by our experiences and expectations in a given context (see the Chapter by M. Gschwind, H. Brettel and I. Rentschler, this volume). At the same time, the visual system is understood to operate as a hard-wired, intrinsic mechanism that leaves little room for expectancy, context, and suchlike. We will therefore evaluate our description against the backdrop of what is known about the workings of the visual cortex, the phenomenology of completion as well as the experimental work in the field.

2 Genealogy of Completion

Perceptual completion is an umbrella term for all phenomena in which incomplete information is perceptually completed. Although the members within this family are related (Pessoa and de Weerd 2003), distinctions can be made based on the nature of the incompleteness they solve.

A particularly well-known example of perceptual completion is the filling-in of the blind spot, the retinal region from which no information can be relayed to the brain. This missing information is perceptually completed with the perceptual properties of its surround (Lettvin 1976; Pessoa et al. 1998; Ramachandran and Gregory 1991). This is a unique type of completion because it remedies incompleteness resulting from the wiring of the retina. In what follows we will concentrate on completion that arises from the presence of objects or shapes in the environment. In these cases anticipated disclosure may be of greater importance.

2.1 Amodal and Modal Completion

There are different ways of completing missing information from the light that reaches the retina and "disclosure anticipation" does not play a role in all of them. A distinction is traditionally made between *amodal* and *modal* completion. Amodal completion was first characterized by the observation that when a moving object disappears in a tube that lies on its trajectory and subsequently reappears, its trajectory nevertheless seems continuous and uniform. Because this perceptual experience is not accompanied by visual sensation of the missing information, it was called amodal (Michotte and Burke 1951; Michotte et al. 1964). This observation generalizes to static images in which the presence of an occluded region is experienced without the sensory qualities of normal vision.

With modal completion, visual experiences of brightness, color or contours arise that are not locally supported by spectral properties of the reflected light. Typical examples of these are the well-known Kanizsa figures, in which a set of arranged cut-out circles conveys the impression that there is a surface connecting them (see Fig. 1). Here, the completed parts are present in consciousness; their sensory properties, such as the increased brightness inside of the illusory boundary, can be commented upon. These perceptions are, however, illusory. They cannot be verified by other senses and are most strongly evoked by pictorial displays. In modal completion the visual system mistakenly treats the stimulus as incomplete; there is no outside source for it.

2.2 Useful and Absurd Completion

Whereas modal completion can be categorized as illusory, amodal completion is systematically supported by the environment. In natural environments, objects rarely end where they cease to be visible and therefore disclosure often follows momentary occlusion. The function of amodal completion may be to correctly anticipate this so that action can be guided accordingly.

To correctly anticipate disclosure can be an important asset for survival. It is therefore not surprising that amodal completion can be demonstrated in a variety of animals ranging from young chicks (Regolin et al. 2004; Regolin and Vallortigara 1995), and rodents (Kanizsa et al. 1993) to primates (Deruelle et al. 2000; Fujita 2001; Yamada et al. 1993). These observations suggest that at least part of the ability to represent occluded objects may be hardwired as the product of evolution.

Perceived completions can be at odds with what we know about the world. Perceptual completion may give rise to very unlikely interpretations in pictorial displays such as the one in Figure 2 (Kanizsa and Gerbino 1982).

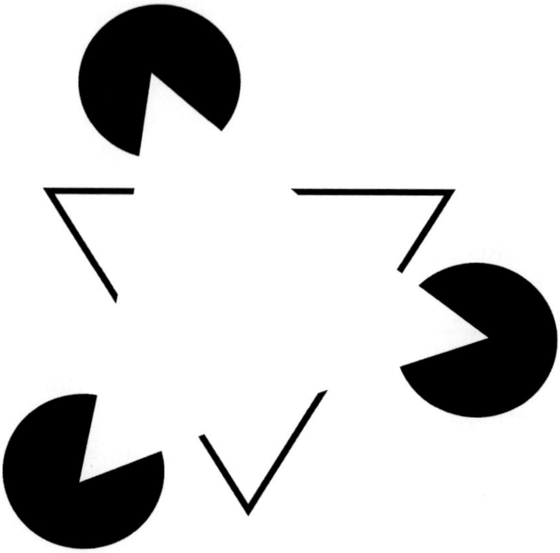

Fig. 1. Modal and amodal completion. Here a triangle seems to partly occlude three circles and another triangle, the first triangle is an example of modal completion, the other figures are amodally completed, after (Kanizsa 1955). We experience this surface as a region of enhanced brightness, compared to its ground. The illusory contrast produces the impression of a distinct, triangular shape on a darker ground. In these figures the completed part displays sensory attributes just like the rest of the figure; both are said to be in the same mode

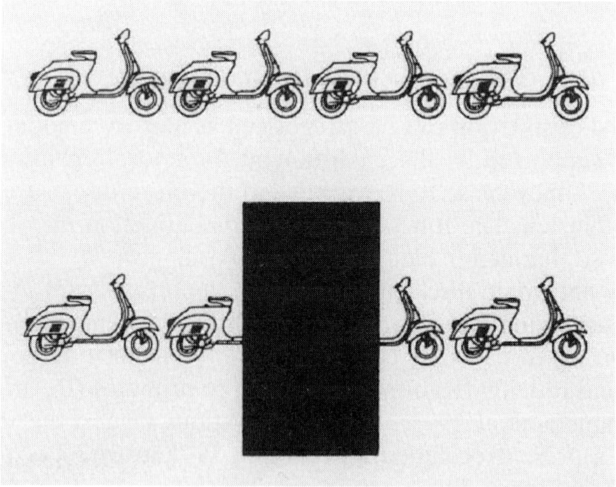

Fig. 2. The power of perceptual processes in amodal completion. Despite the favorable context, a very unlikely scooter is seen to continue behind the occluder (from Kanizsa and Gerbino 1982)

We must, therefore, be careful not to confuse the notion that the visual system anticipates disclosure of an object with the view that this anticipation is based on veridical reasoning. Kanizsa's illustration tells us that completion can only be explained in terms of expectancies in a very narrow sense. The knowledge available to the visual system in anticipating a complete figure is limited, almost certainly excluding certain aspects of the semantics of the picture. Evolutionary explanations for these restrictions may involve speed constraints; perception needs to be fast as well as reliable. As a result, the way these solutions are obtained is likely to be hardwired in the architecture of the visual system.

These observations may seem to further constrain the role of experience in completion. The observed properties of the visual cortex, however, suggest that context can strongly modulate how occluded figures are processed (Albright 1995; Assad and Maunsell 1995). Even early visual areas may be reconfigured by experience and the current state of the perceiver (Ahissar et al. 1992; Ahissar and Hochstein 1993). The processing of image features on the neural level is sensitive to context (Albright and Stoner 2002), and so are the response properties of primary visual cortex (Lee et al. 2002). A crucial question for our understanding of completion, therefore, is, in what way it is sensitive to context. But before we turn to the process of completion a description is needed of what completion does.

3 Completion Creates Wholes

Partly occluded figures are represented as wholes (Gerbino and Salmaso 1987). When completion is based on an anticipation of an object, we cannot determine the nature of the complete object directly from the visible stimulus properties. Instead, we must understand it from the properties of the whole object.

The first evidence for the role of whole object structure came from the effect of Goodness on completion (Buffart and Leeuwenberg 1981; Buffart et al. 1983). In a matching experiment, a geometrical shape was partly occluded and two or more un-occluded alternatives were presented, from which one had to be chosen. Participants preferred completions that were based on measures of the figural Goodness of the completed figure. The preference for Goodness is generally considered as intrinsic to the perceptual system (Leeuwenberg 1971; van der Helm and Leeuwenberg 1996). This squares with the notion that an innate perceptual architecture can completely account for amodal completion.

3.1 Local and Global Factors

Goodness factors that may contribute to completion include symmetry, similarity, proximity, and good volume continuation or complete mergeability (Tse 1999a, b). We will distinguish between local and global factors. An example of a local factor is good continuation. Good continuation can be achieved by interpolating contours behind an occluder. This can sometimes be sufficient to explain amodal completion (Kanizsa and Gerbino 1982; Kellman and Shipley 1991; Shipley and Kellman 2003). Local completions of line segments, curves and edges can be processed in primary visual cortex as the continuation of lines at T-junctions and grouping based on proximity of parts (Dresp and Grossberg 1997; Field et al. 1993; Kovacs and Julesz 1993).

Local accounts can deal with only a subset of completion phenomena. Completions can also be derived from global properties of the occluded figure (Buffart and Leeuwenberg 1981; Buffart et al. 1983; Sekuler 1994; van Lier et al. 1995). Global effects in occlusion are related to the Gestalt principles of symmetry and closure.

Global representations can also be arrived at in primary visual areas. These areas display sensitivity to non-local properties of the visual structure quickly after stimulus onset (Altmann et al. 2003; Kamitani and Shimojo 2004; Nikolaev and van Leeuwen 2004). Global processing may be realized through lateral interactions in the visual cortex, in particular in area V2 (Peterhans and von der Heydt 1989).

In cases such as in Figure 3, both local and global completions are possible for the same pattern (Buffart et al. 1983). Such occluded figures prime both of their alternative completion interpretations (van Lier et al. 1995). This result suggests the possibility that both alternative completions were made when the occlusion was presented. To evaluate this possibility, we consider completion as a process.

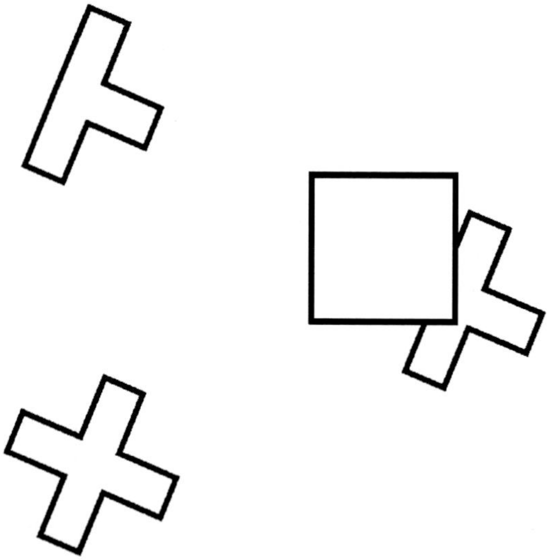

FIG. 3. Local and global completions. The occluded figure on the left can have a local completion based on good continuation (top-right), and a global interpretation based on optimized symmetry (bottom-right)

3.2 The Process of Completion

Amodal completion is a process of variable duration (Guttman et al. 2003; Murray et al. 2001). The smallest estimate of completion time to date is 75 ms, obtained in a psychophysical task in which participants judged whether a partly occluded square was higher than its width (Murray et al. 2001). The largest estimate comes from an experiment in which participants viewed a prime (circles or squares) that was either occluded by another object, or shown in plain view. Immediately thereafter, they judged whether two simultaneously presented figures were same or different. Occluded figures facilitated responses to their whole, un-occluded counterpart only when they were presented longer than 200 ms (Sekuler and Palmer 1992).

Two factors systematically influence completion time. The first is the amount of depth cues present in the stimulus display (Bruno et al. 1997). The second is the size of the occluded parts; the larger occlusions take more time to be completed (Rensink and Enns 1998; Shore and Enns 1997).

Size-dependency of completion times has been confirmed by a visual search study (Rauschenberger and Yantis 2001). Subjects searched for truncated figures that could lay adjacent to squares. Search for truncated figures would have been easy, had the perceiver been able to suppress their completion (He and Nakayama 1992; Rensink and Enns 1995). This search was inefficient for long presentation times; in that case the figures were amodally completed. The results therefore suggest that the completion process is mandatory. The presentation times needed for this inefficiency to arise depended on the amount of apparent occlusion. When Rauschenberger and Yantis (2001) masked the search display briefly after presentation, search became efficient again. They suggested that the completion process did not develop far enough to preempt access to a mosaic-like interpretation of the target figure.

Can we distinguish stages in the process that lead to completion? Sekuler and Palmer (1992) only observed priming effects for whole, completed figures when the occluded prime was presented longer than 200 ms. For brief presentations (50 ms), however, the occluded shapes primed truncated figures. This result suggests that there is a stage prior to completion, in which an occluded figure is transiently represented as a 2D mosaic.

The two-stage view of completion was supported by results from a shape-discrimination task (Ringach and Shapley 1996). The existence of a mosaic stage cannot always be confirmed, however. In experiments with enhanced 3D cues, the mosaic-stage was either absent or passed very quickly (Bruno et al. 1997). It may still be, however, that a mosaic representation is computed in parallel, at least in part, with completion. At present, the existence of a separate the mosaic-stage is controversial (Plomp et al. 2006).

What is important for our current discussion is that multiple interpretations of the same figure can be generated. The time it takes to arrive at an interpretation suggests that the process goes beyond elementary visual operations; the effects of higher-level influence on primary visual areas start at approximately 100 ms

(Vanni et al. 2001; Zipser et al. 1996). Given the variability in completion times we may consider the process a flexible and active one. This means it is leaving enough room for contextual modulation, in line with the description of completion as an anticipated disclosure serving future action. In the following we will specify in what ways amodal completion may depend on the sate of the environment and the state of the observer.

4 The Role of Context

Context may be defined as the temporal and spatial circumstance in which perception occurs, encompassing the history of the perceiver as well as the immediately available information in the environment. The notion of context thus emphasizes the current state of the perceiver and its environment.

4.1 Spatial Context

Dinnerstein and Wertheimer (1957) provided an early demonstration of the role of spatial context on completion. In Figure 4, the surrounding context seriously attenuates the completion of a partly occluded square. The effect is based on the rivaling symmetry of the four L-shaped figures.

In a visual search study (Rauschenberger et al. 2004), the authors take a new look at some of their earlier findings (Rauschenberger and Yantis 2001). In the earlier study the inefficient search for truncated shapes was attributed to mandatory completion that resulted in occlusion interpretations of these shapes. In that

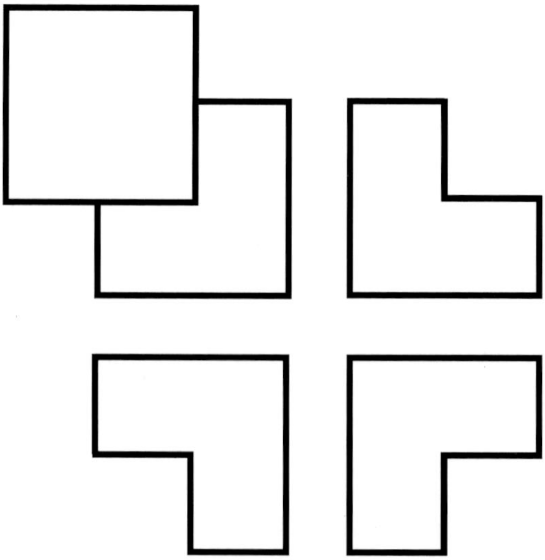

Fig. 4. The surrounding context can weaken amodal completion (Dinnerstein and Wertheimer 1957)

study, however, the figures surrounding the target figure were compatible with occlusion interpretations. These surrounding non-targets may have induced the completion interpretations of the targets (Peterson and Hochberg 1983). Rauschenberger et al. (2004) demonstrated that non-targets in the search display do influence the interpretation of the targeted figure. In this study search for a partly occluded circle was impeded when the surrounding figures were notched circles, as in Figure 4. This provides evidence for the influence of spatial context on completion.

4.2 Temporal Context

The effect of temporal context is that of prior exposure. The effect of prior exposure on completion was demonstrated by presenting subjects with vertical bars of different lengths that were subsequently occluded (Joseph and Nakayama 1999). After occlusion, the un-occluded parts of the bars could induce either vertical or horizontal motion. Although the occlusion displays were the same for short and long occluded bars, perceiver's history affected how the bars were completed, resulting in differences in the perceived motion.

Temporal context may also act to override the initial interpretation of occluded figures. Stimulus configurations that at first viewing do not give rise to amodal completion may do so after a congruent interpretation has been presented to suggest this. In this way, even disparate fragments can appear amodally completed behind an occluder (Zemel et al. 2002).

Repeated prior exposure leads to increased familiarity. We previously showed how familiarity affects completion (Plomp et al. 2004). In two experiments we measured eye movements of subjects who were engaged in a visual search for target figures that could sometimes be partly occluded. Gaze durations on these figures were taken as a measure of the time needed to complete them. The familiarity of the target figure was found to be the factor that determined gaze durations when they were partly occluded; familiar completions were performed faster than unfamiliar ones. These results can be interpreted as long-term effects of repeated exposure on completion.

4.3 Context Affects Visual Processes

The above demonstrations leave open the question of *how* context effects arise in completion, whether they affect visual processes or post-perceptual ones. To answer this question we looked at the effects of prior exposure in an extended primed-matching paradigm (Plomp 2005). In these experiments, subjects were presented two primes before the test pair; one was an occluded figure and the other could be compatible or incompatible with it. These figures were called compatible when they resembled the local or global completion of the occluded figure, or a mosaic interpretation of it.

We studied the combined effect of the primes on the RT to the task. The results showed a super-additive interaction between the two primes; when the second prime was an occluded figure and the first one was compatible with it their combined effect was larger than the effect of each of the two primes separately. This indicates that the interpretation of the second, occluded figure was biased by the preceding figure. All three interpretations showed such an effect, suggesting that both completion interpretations and the mosaic one are present in the completion process. The interaction between the two primes demonstrates the effect of prior exposure on the processing of an occlusion. The effect was restricted to short presentations of the occluded figure and was dependent on the temporal order of the two figures. The results showed that preceding context serves to bias possible interpretations *during* the process of completion. However, as soon as the completion process is finished, the representation became immune to the effect of prior exposure. Thus it is unlikely that post-perceptual decision processes are responsible for the effects of prior context.

5 Conclusions

The problem of amodal completion can be stated as how the brain arrives at an actable interpretation of the current 3D environment from its limited 2D projection. The interpretation not only goes beyond the available information, but often also beyond what can be inferred based on simple interpolation processes. Multiple completions of the same figures are made in parallel, although they may finish at different rates. The process of completion can be characterized as a flexible and active one: context influences on this process play a role prior to completion, generating expectancy, during completion, facilitating certain completions, as well as afterwards, in deciding which of possible alternatives is preferred. Understanding its flexibility will be of crucial importance if the process of occlusion is to be disclosed.

References

Ahissar E, Vaadia E, Ahissar M, Bergman H, Arieli A, Abeles M (1992) Dependence of cortical plasticity on correlated activity of single neurons and on behavioral context. Science 257:1412–1415

Ahissar M, Hochstein S (1993) Attentional control of early perceptual learning. Proc Natl Acad Sci U S A 90:5718–5722

Albright TD (1995) My most true mind thus makes mine eye untrue. Trends Neurosci 18:331–333

Albright TD, Stoner GR (2002) Contextual influences on visual processing. Annu Rev Neurosci 25:339–379

Altmann CF, Bülthoff HH, Kourtzi Z (2003) Perceptual organization of local elements into global shapes in the human visual cortex. Curr Biol 13:342–349

Assad JA, Maunsell JH (1995) Neuronal correlates of inferred motion in primate posterior parietal cortex. Nature 373:518–521

Bruno N, Bertamini M, Domini F (1997) Amodal completion of partly occluded surfaces: is there a mosaic stage? J Exp Psychol Hum Percept Perform 23:1412–1426

Buffart H, Leeuwenberg E (1981) Coding theory of visual pattern completion. J Exp Psychol Hum Percept Perform 7:241–274
Buffart H, Leeuwenberg E, Restle F (1983) Analysis of ambiguity in visual pattern completion. J Exp Psychol Hum Percept Perform 9:980–1000
Deruelle C, Barbet I, Depy D, Fagot J (2000) Perception of partly occluded figures by baboons (Papio papio). Perception 29:1483–1497
Dinnerstein D, Wertheimer M (1957) Some determinants of phenomenal overlapping. Am J Psychol 70:21–37
Dresp B, Grossberg S (1997) Contour integration across polarities and spatial gaps: from local contrast filtering to global grouping. Vision Res 37:913–924
Field DJ, Hayes A, Hess RF (1993) Contour integration by the human visual system: evidence for a local "association field". Vision Res 33:173–193
Fujita K (2001) Perceptual completion in rhesus monkeys (Macaca mulatta) and pigeons (Columbia livia). Percept Psychophys 63:115–125
Gerbino W, Salmaso D (1987) The effect of a modal completion on visual matching. Acta Psychol (Amst) 65:25–46
Gibson JJ (1972) A theory of direct visual perception. In: Royce JR, Rozeboom WW (Eds) The psychology of knowing. New York, New York, pp 215–240
Guttman SE, Sekuler AB, Kellman PJ (2003) Temporal variations in visual completion: a reflection of spatial limits? J Exp Psychol Hum Percept Perform 29:1211–1227
He ZJ, Nakayama K (1992) Surfaces versus features in visual search. Nature 359: 231–233
Joseph JS, Nakayama K (1999) Amodal representation depends on the object seen before partial occlusion. Vision Res 39:283–292
Kamitani Y, Shimojo S (2004) Global yet early processing of visual surfaces. In: Chalupa LM, Werner JS (Eds) The visual neurosciences. MIT Press, Cambridge, Massachusetts, pp 1129–1138
Kanizsa G (1955) Margini quasi-percettivi in campi con stimolazione omogenea. Riv Psicol 49:7–30
Kanizsa G, Gerbino W (1982) Amodal completion: seeing or thinking? In: Beck J (Ed) Organisation and representation in perception. Hillsdale New Jersey, Hillsdale New Jersey, pp 167–190
Kanizsa G, Renzi P, Conte S, Compostela C, Guerani L (1993) Amodal completion in mouse vision. Perception 22:713–721
Kellman PJ, Shipley TF (1991) A theory of visual interpolation in object perception. Cognit Psychol 23:141–221
Kovacs I, Julesz B (1993) A closed curve is much more than an incomplete one: effect of closure in figure-ground segmentation. Proc Natl Acad Sci USA 90:7495–7497
Lee TS, Yang CF, Romero RD, Mumford D (2002) Neural activity in early visual cortex reflects behavioral experience and higher-order perceptual saliency. Nat Neurosci 5:589–597
Leeuwenberg EL (1971) A perceptual coding language for visual and auditory patterns. Am J Psychol 84:307–349
Lettvin J (1976) On seeing sidelong. Sciences 16:10–20
Michotte A, Burke L (1951) Une nouvelle enigme dans la psychologie de la perception: le "donne amodal" dans l'experience sensorielle. In: Proceedings of the 13th international congress of psychology, pp 179–180
Michotte A, Thinès G, Crabbé G (1964) Les Compléments Amodaux des Structures Perceptives. Louvain: Institute de psychologie de l'unversité de Louvain

Murray RF, Sekuler AB, Bennett PJ (2001) Time course of amodal completion revealed by a shape discrimination task. Psychon Bull Rev 8:713–720

Nikolaev AR, van Leeuwen C (2004) Flexibility in spatial and non-spatial feature grouping: an event-related potentials study. Cogn Brain Res 22:13–25

Pessoa L, de Weerd P (2003) Filling-in: from perceptual completion to cortical reorganization. New York, New York

Pessoa L, Thompson E, Noe A (1998) Finding out about filling-in: a guide to perceptual completion for visual science and the philosophy of perception. Behav Brain Sci 21:723–748; discussion 748–802

Peterhans E, von der Heydt R (1989) Mechanisms of contour perception in monkey visual cortex. II. Contours bridging gaps. J Neurosci 9:1749–1763

Peterson MA, Hochberg J (1983) Opposed-set measurement procedure: a quantitative analysis of the role of local cues and intention in form perception. J Exp Psychol Hum Percept Perform 9:183–193

Plomp G (2005) Amodal completion of occluded figures: what context uncovers. University of Sunderland, Sunderland

Plomp G, Liu L, van Leeuwen C, Ioannides AA (2006) The "mosaic stage" in amodal completion as characterized by magnetoencephalography responses. J Cogn Neurosci 18(8):1394–1405

Plomp G, Nakatani C, Bonnardel V, van Leeuwen C (2004) Amodal completion as reflected by gaze durations. Perception 33:1185–1200

Ramachandran VS, Gregory RL (1991) Perceptual filling in of artificially induced scotomas in human vision. Nature 350:699–702

Rauschenberger R, Yantis S (2001) Masking unveils pre-amodal completion representation in visual search. Nature 410:369–372

Rauschenberger R, Peterson MA, Mosca F, Bruno N (2004) Amodal completion in visual search: preemption or context effects? Psychol Sci 15:351–355

Regolin L, Vallortigara G (1995) Perception of partly occluded objects by young chicks. Percept Psychophys 57:971–976

Regolin L, Marconato F, Vallortigara G (2004) Hemispheric differences in the recognition of partly occluded objects by newly hatched domestic chicks (Gallus gallus). Anim Cogn 7:162–170

Rensink RA, Enns JT (1995) Preemption effects in visual search: evidence for low-level grouping. Psychol Rev 102:101–130

Rensink RA, Enns JT (1998) Early completion of occluded objects. Vision Res 38:2489–2505

Ringach DL, Shapley R (1996) Spatial and temporal properties of illusory contours and amodal boundary completion. Vision Res 36:3037–3050

Sekuler AB (1994) Local and global minima in visual completion: effects of symmetry and orientation. Perception 23:529–545

Sekuler AB, Palmer SE (1992) Perception of partly occluded objects: a microgenetic analysis. J Exp Psychol Gen 21:95–111

Shipley TF, Kellman PJ (2003) Boundary completion in illusory contours: interpolation or extrapolation? Perception 32:985–999

Shore DI, Enns JT (1997) Shape completion time depends on the size of the occluded region. J Exp Psychol Hum Percept Perform 23:980–998

Tse PU (1999a) Complete mergeability and amodal completion. Acta Psychol (Amst) 102:165–201

Tse PU (1999b) Volume completion. Cognit Psychol 39:37–68
van der Helm PA, Leeuwenberg EL (1996) Goodness of visual regularities: a nontransformational approach. Psychol Rev 103:429–456
van Lier RJ, Leeuwenberg EL, Van der Helm PA (1995) Multiple completions primed by occlusion patterns. Perception 24:727–740
Vanni S, Tanskanen T, Seppa M, Uutela K, Hari R (2001) Coinciding early activation of the human primary visual cortex and anteromedial cuneus. Proc Natl Acad Sci U S A 98:2776–2780
Yamada W, Fujita N, Masuda N (1993) Amodal completion as another perception of color-spreading stimuli. Percept Mot Skills 76:1027–1033
Zemel RS, Behrmann M, Mozer MC (2002) Experience-dependent perceptual grouping and object-based attention. J Exp Psychol Hum Percept Perform 28:202–217
Zipser K, Lamme VA, Schiller PH (1996) Contextual modulation in primary visual cortex. J Neurosci 16:7376–7389

2 Functional MRI Evidence for Neural Plasticity at Early Stages of Visual Processing in Humans

SOPHIE SCHWARTZ[1,2]

1 Introduction

In everyday life, our visual system is continuously flooded with information from the environment. However, the visual system has a limited processing capacity. Hence, we perceive only a fraction of this information, which typically forms the objects of our visual experience. In other words, because of the limited processing resources, information or objects that are simultaneously present in the visual field will compete for neural representation. How does our visual system select what is relevant to us at any given time and in any given context stands as a fundamental question in visual neurosciences.

In recent years, functional magnetic resonance imaging (fMRI) has proven very useful to study visual selection and competition in the human brain (e.g., Kastner and Ungerleider 2000). Here I review three fMRI studies showing that perceptual learning and voluntary attention can bias visual selection and modulate neuronal response in adult human visual cortex. By enhancing the visual processing of relevant information and reducing the processing of ignored stimuli, both learning and attention shape the landscape of our present and future visual experiences.

The studies reported here indicate that substantial neural plasticity may occur at the earliest cortical stage of visual processing, i.e., within the primary visual cortex (V1). More generally, these recent functional neuroimaging data indicate that the influence of learning and attention on early vision is mediated by subtle interactions between excitatory and inhibitory neural mechanisms. They also exemplify the successful integration of behavioral and brain imaging data, shedding new light on the ever-changing and adaptive nature of our brains and minds.

[1]Neurology and Imaging of Cognition, Department of Neurosciences, University Medical Center, Michel-Servet 1, 1211 Geneva, Switzerland
[2]Neurology Clinic, Geneva University Hospital, Micheli-Du-Crest 24, 1211 Geneva, Switzerland

2 Perceptual Learning Modifies Long-Term Retinotopic Response in Primary Visual Cortex

One fundamental property of brain systems is to adapt their functions in response to environmental changes. Recent neurophysiological studies in adult monkeys show that such experience-dependent neural changes may occur as early as in the primary visual cortex (V1), where single-neuron responses can be permanently affected by exposure to novel visual stimuli (for reviews, see Gilbert et al. 2001; Tsodyks and Gilbert 2004). Also suggestive of V1 contribution in visual learning, psychophysical improvements after visual discrimination learning in adult humans are often restricted to the trained stimulus configuration, such as the orientation of stimulus elements, location in the visual field, and training of the eye (Karni and Sagi 1991; Crist et al. 2001). Experience-dependent changes might thus take place at early processing stages in the visual system where eye-specificity, orientation information, and retinotopic location of visual inputs are mapped with the highest resolution. Based on these cellular and behavioral data, we designed an experiment to test for learning-related changes in V1 of adult humans.

Using fMRI, we measured neural activity 24 hours after participants were intensively trained in visual texture discrimination, when the task was performed with one eye and within one visual quadrant (Fig. 1c; Schwartz et al. 2002; Walker et al. 2005). In this task, participants were asked to determine the orientation of a peripheral target-texture, while simultaneously monitoring the identity of a central letter (Fig. 1a). Performance is known to improve only for the trained location and the trained eye (Karni and Sagi 1991). As targets were always presented in the upper-left visual quadrant, the corresponding response in the visual cortex occurred in the same retinotopic regions of visual cortex, and varied only as a function of the learning status of the tested eye (i.e., trained or untrained). We could thus directly compare brain activity associated with performing the task with either the trained or the untrained eye (Fig. 1b), and test for any learning-dependent changes in the BOLD (blood oxygenation level-dependent) response 24-hours after training.

Individual performance assessed 24-hours after training confirmed a selective improvement at discriminating the peripheral target with the trained eye as compared to the untrained eye. Whole-brain fMRI data were analyzed using SPM (http://www.fil.ion.ucl.ac.uk/spm/). At the group level, the comparison "trained > untrained eye" demonstrated a single region of increased activity located in the lower bank of the right calcarine sulcus, which corresponded precisely to the retinotopic projection of the stimulated upper left quadrant onto the V1 (Fig. 1d, e). There was no other significant fMRI increase detected throughout the brain for either the same comparison or for the reverse comparison (untrained > trained condition).

Retinotopic increase in BOLD response 24 hours after visual learning provides important empirical support for recent theoretical models in which perceptual

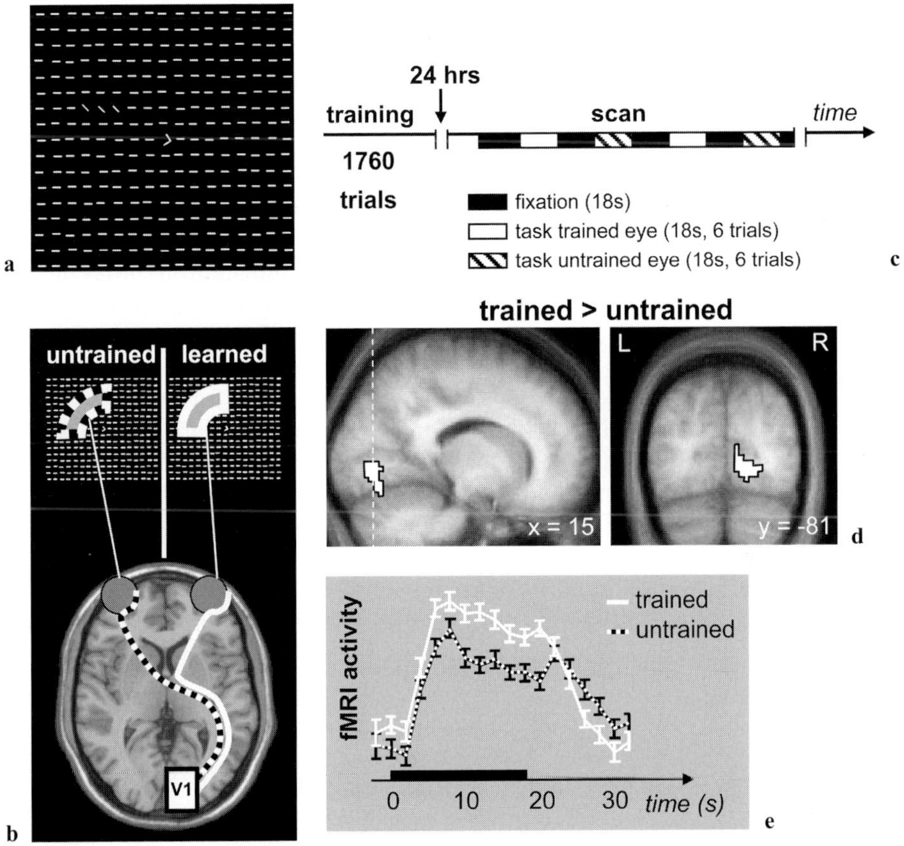

FIG. 1. **a** Stimulus display. On each trial, participants had to identify a rotated central letter (L or T) and the orientation of a target-texture, i.e., three diagonal elements next to each otehr (horizontal texture) or on top of each other (vertical texture). **b** Stimuli were always shown in the upper left quadrant of the visual field, i.e., were projected to the lower part of the right visual cortex both when seen with the trained eye *(solid white line)* or the untrained eye *(dashed line)*. **c** Training was performed 24-hours before scanning, using one eye only. During scanning, 24 fixation blocks *(black)*, each followed by one task block of 6 visual discrimination trials, were presented to the trained *(white)* and untrained *(shaded)* eye in alternation. **d** Increased MRI signal in the lower bank of the right calcarine sulcus (right lingual gyrus) in the learned condition (trained eye) as compared to the new condition (untrained eye), within the retinotopic projection of the upper-left quadrant. Group results superimposed onto the mean normalized anatomical brain of participants. **e** Group-averaged MRI response in the right occipital peak, for the trained *(solid white line)* and untrained *(dashed line)* conditions, demonstrating an enhanced response for the trained eye as compared to the untrained eye. Adapted from Schwartz et al. (2002)

learning favors inhibitory activity in the visual cortex in order to increase the discrimination of trained targets from background flankers (Herzog and Fahle 1998; Tsodyks and Gilbert 2004). Furthermore, learning-dependent increases in V1 responsiveness to a trained visual configuration may result from changing contextual influences exerted by stimuli outside the classical receptive field (Gilbert et al. 2001; Grossberg and Williamson 2001; Hupe et al. 2001). In the present study, contextual tuning would enhance the visual segregation of contiguous diagonal lines from a homogeneous background of horizontal lines (Fig. 1a). The recruitment of larger assemblies of interconnected neurons after learning could then produce a higher total neural response to the target-texture, associated with increased, regionally-specific BOLD response (Logothetis et al. 2001).

In conclusion, this first fMRI study provides evidence for retinotopically-specific increased activity in V1 after training on a fine discrimination task performed within one visual-quadrant. Our findings thus demonstrate that perceptual experience may trigger lasting functional reorganization within the early visual cortex of adult humans (see also Furmanski et al. 2004).

However, spatial attention can also influence early visual responses in a retinotopic way (e.g., Tootell et al. 1998; Somers et al. 1999). (Note that, in the first study reported above, all stimuli were shown within the upper-left quadrant for both the trained and untrained conditions, thus controlling for any effect of spatial attention.) Much like visual discrimination learning, attention might provide important constraints on the processing of visual inputs, reflected by a nonhomogeneous distribution of neural activity in retinotopically-organized visual cortices. In the next section, we report a second study that tested whether attention involves an interaction between excitatory and inhibitory influences that would strengthen the processing of information at the attended location but suppress the processing of information from areas surrounding the focus of attention.

3 Attention Modulates Neural Activity for Task-Irrelevant Peripheral Visual Stimuli

Previous fMRI studies have shown that attention can enhance the fMRI signal at early cortical stages of visual processing, including the primary visual cortex (Tootell et al. 1998; Somers et al. 1999). Recent theories of attention suggest, however, that spatial attention does not only enhances processing at attended locations but may also selectively suppress processing at non-attended locations (see Lavie and Tsal 1994; Lavie 2005). Increased "attentional load" at central fixation (e.g., a more difficult task at fixation) would thus cause less processing of the peripheral field (and hence less interference from distractors). High attentional load for central targets might therefore lead to a reduction of neural activity triggered by unattended inputs (see Rees et al. 1997; Smith et al. 2000; Pinsk et al. 2004). However, the topography of selective suppression remains controversial. Increased attentional load at fixation might either produce so-called "tunnel vision" (more eccentric locations being the most affected; Ikeda and Takeuchi 1975; Chan and Courtney 1998), or an effect of "surround suppression" (locations closer to the attended location being the most affected; Bahcall and Kowler 1999; Plainis et al. 2001), or a uniform reduction in peripheral processing across all eccentricities in the field (Holmes et al. 1977; Williams 1984).

To formally test these hypotheses, we designed an fMRI experiment in which we varied attentional load (low or high load) in a central task, while presenting flickering checkerboards as task-irrelevant stimuli in the peripheral visual field (Fig. 2a; Schwartz et al. 2004). During the main fMRI experiment, a group of healthy participants performed a visual detection task on a continuous rapid stream of colored T-shaped stimuli shown with different orientations (upright or upside-down) at fixation (Fig. 2b). Participants were required to monitor for the occurrence of infrequent pre-specified targets: during the easy/low-load condition, the targets were red Ts irrespective of their orientation; during the difficult/high-load condition, the targets were any upright yellow T or upside-down green T (both types of conjunction targets had to be monitored throughout this task).

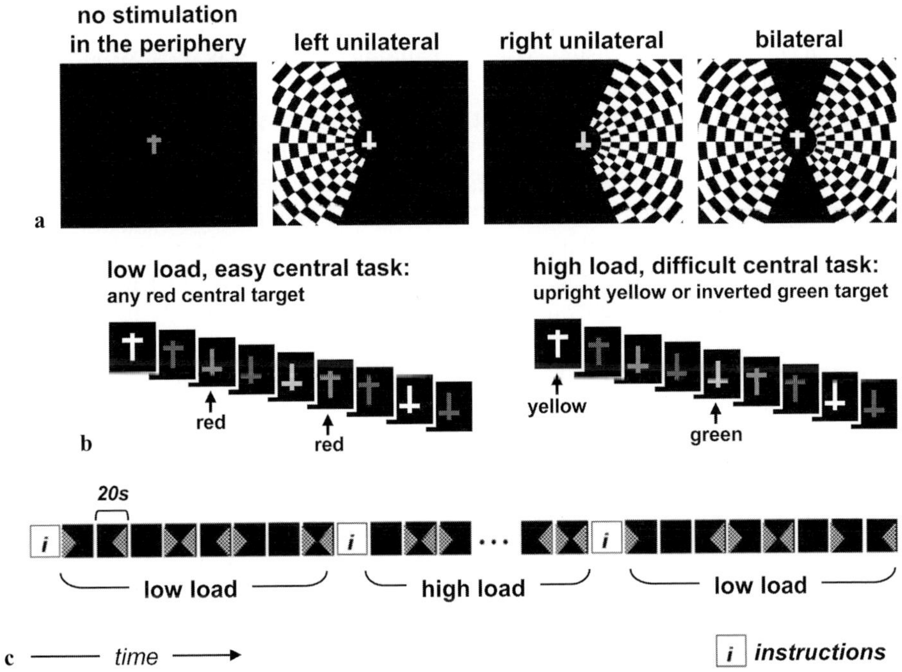

FIG. 2. **a** The four visual conditions included blocks of 20 sec with flickering checkerboards presented to either the right, the left, or to both hemifields or none. **b** A rapid continuous stream of colored T shapes appeared at central fixation during all conditions. In the low-load task, participants had to detect any red shape; in the high-load task, they had to detect yellow upright or green inverted Ts. **c** Blocks with irrelevant checkerboard stimulation in either hemifield (none, unilateral left or right, and bilateral) alternated during both task conditions. Adapted from Schwartz et al. (2004)

The central target stream was shown continuously but presented either alone, or accompanied by peripheral flickering checkerboards that could appear in either the right, the left, or both visual fields, in randomly ordered blocks of 20s-duration each (Fig. 2a, c). Importantly the central stimuli were equivalent in all respects across the two task conditions (Fig. 2b); only the task instructions distinguished the high-load and low-load conditions for the central task. Each task was performed twice during 160-sec periods, each separated by a 20-sec display that presented instructions for the next task (high-load or low-load; Fig. 2c).

A standard fMRI retinotopy protocol followed the visual-load experiment during the same scanning session (Sereno et al. 1995; DeYoe et al. 1996; Engel et al. 1997; Fig. 3a, b). The visual stimuli used there (wedge and annulus) covered the same extent of the visual field that had been stimulated by the full, task-irrelevant peripheral checkerboards in the load experiment.

As assessed behaviorally (reaction-times and hit rates), the task was indeed harder for the high- than the low-load condition in all participants, confirming that central attentional load was successfully varied by our task assignments. The fMRI results for functionally-defined retinotopic areas mapped in 6 of the participants (12 hemispheres) revealed a reduction of cortical activation for the peripheral visual stimuli during higher attentional load at central fixation which occurred throughout the visual cortex (including V1) but was most pronounced in higher-level extrastriate areas (Fig. 3c).

We also delimited separate "eccentricity bins" for the peripheral visual field within individual cortical areas (Fig. 3b) to test whether increased attentional load in the central task might differentially affect visual field locations of differing eccentricity. Indeed, higher load in the central task produced a larger reduction of the response to a contralateral stimulus for voxels representing the "inner" (2–8°) visual field than for those representing the "outer" (8–14°) field further away from the attended central stream (main effect of eccentricity for low-load minus high-load conditions, $P < 0.05$ for both unilateral/contralateral and bilateral stimulation but there was no eccentricity effect for ipsilateral or absent peripheral stimulation; Fig. 3d).

These data therefore suggest that when more attentional capacity is allocated at central fixation, cortical activation for task-irrelevant peripheral stimulation is reduced primarily for the representation of adjacent central portions of the visual field (consistent with "surround-suppression" proposals, see Bahcall and Kowler 1999; Plainis et al. 2001), but less so for the more eccentric locations that are further away from the central stimuli. Moreover, suppressive effects of a high central load were larger in the presence of checkerboard stimuli in the contralateral hemifield across all visual areas, but particularly for V1. This suggests that load-related reduction in early visual areas may primarily affect stimulus-driven responses to peripheral distractors, as predicted by research on perceptual load (Lavie and Tsal 1994; Lavie 2005; O'Connor et al. 2002).

While attention can modulate activity in early visual areas in a retinotopic manner for stimuli at attended locations (e.g., Brefczynski and DeYoe 1999; Gandhi et al. 1999) or even in the absence of stimuli (Kastner et al. 1999), our

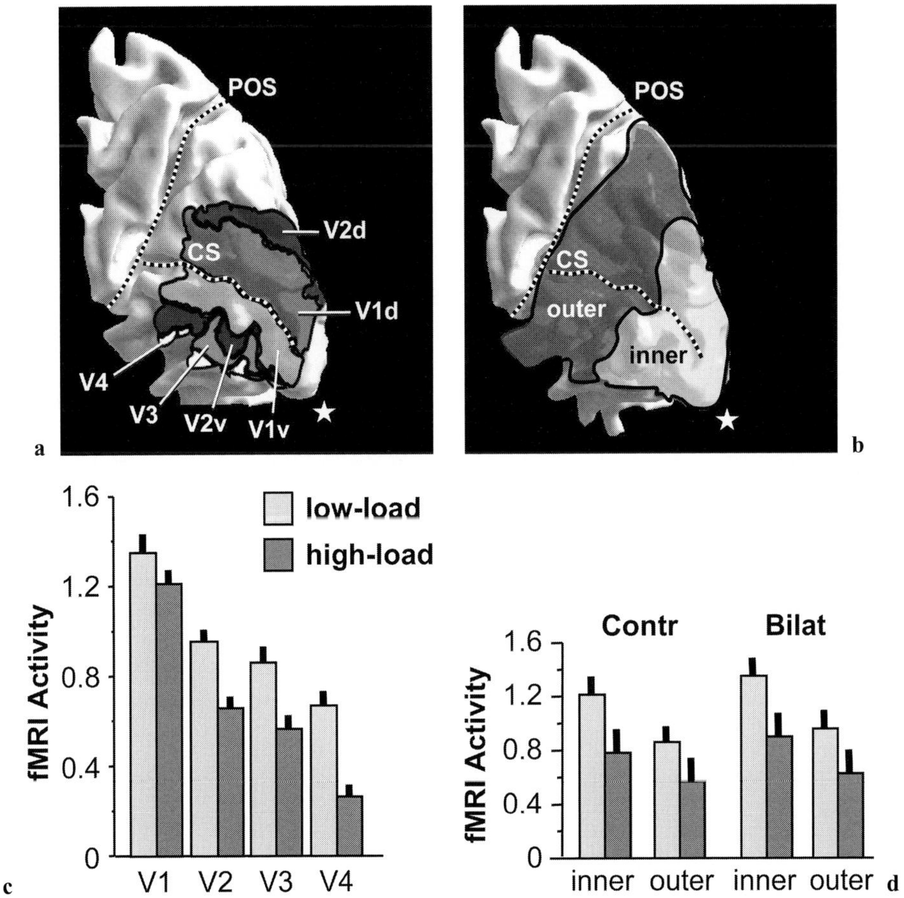

FIG. 3. **a** Three-dimensional reconstruction of the medial occipital cortex of one participant showing voxels assigned to distinct visual areas (V1, V2, V3, ventral V4) that resulted from retinotopic mapping procedure (CS = calcarine sulcus; POS = parieto-occipital sulcus; asterisk = projection of the foveal region). **b** Eccentricity map for the same participant showing voxels corresponding to the inner and outer (peripheral) regions of the visual field. **c** Activity in visual areas from 12 hemispheres of six participants delimited by the retinotopic mapping procedure. Mean MRI activity (+/– SE) for each area under low (light grey) and high (dark grey) task load -averaged over these conditions with checkerboards in the contralateral hemifield showing a progressive effect of load from V1 to V4. **d** Effects of central load at central and more peripheral locations in the retinotopic cortex. Mean MRI activity (+/– SE) in the cortex representing the "inner" (~2–8°) or "outer" (~8–14°) parts of the central visual field (pooled across all areas). During contralateral and bilateral stimulation, higher load reduced neural responses at inner locations more than at outer locations. This pattern was found in each visual area. Adapted from Schwartz et al. (2004)

findings firmly establish that attention can also affect neural activity in the visual cortex corresponding to stimulation at irrelevant locations. This second fMRI study thus provides new insight into the top-down influence on the processing of both attended and unattended visual information. The goal of the experiments described in the next section is to refine our understanding of how such attentional processes at encoding might also affect subsequent memory for visual information.

4 Long-Term Modulation of Memory and Neural Activity for Previously Attended and Ignored Stimuli

While inattention to visual stimuli at the encoding phase can abolish later recognition in direct explicit tests (Rock and Guttman 1981), residual visual processing still occurs without attention, as demonstrated by indirect tests such as repetition priming, word-stem completion, or degraded picture identification (Parkin et al. 1990; Szymanski and MacLeod 1996; Merikle et al. 2001). Our previous work also demonstrated that neglect in patients with parietal damage and spatial-attention deficits may exhibit delayed priming effects for objects initially presented on the affected side, even when these were not consciously reported at exposure nor explicitly remembered (Vuilleumier et al. 2001; Vuilleumier et al. 2002b). This suggests that some degree of processing can still take place for unattended visual objects, despite the absence of explicit memory.

Attention can selectively privilege the visual processing of some stimuli and suppress other, irrelevant stimuli, when these are positioned at separate spatial locations (see above), but also when both attended and unattended stimuli overlap at the same location (O'Craven et al. 1999; Rees et al. 1999). Such modulation may occur at many stages along the visual pathways, including the primary cortex, but is typically more pronounced at higher levels (Kastner and Ungerleider 2000; Driver and Frackowiak 2001). Conversely, it has been proposed that visual processing of ignored stimuli might be restricted to early perceptual stages in the visual system where objects are coded in a view-specific rather than view-independent manner (Grill-Spector et al. 1999; Vuilleumier et al. 2002a), although semantic priming effects may still occur for unattended visual stimuli (Merikle et al. 2001).

Here we used pairs of overlapping line-drawings of objects as shown in Figure 4a, and asked participants to selectively attend to the objects drawn in one color and not to those in the other color (see Vuilleumier et al. 2005; Yantis and Serences 2003). We thus created a condition where attention would select among stimuli presented simultaneously at the exact same retinal location. This allowed us to subsequently assess the fMRI signal as well as explicit and implicit memory traces, for both previously attended and ignored objects.

After the first exposure to the overlapping objects (study phase), one group of participants was given a surprise memory test, in which they were shown one object at a time (previously attended or ignored objects, either shown in original or mirror view, plus new objects; Fig. 4b). Participants were asked to judge explicitly whether they had already seen the object during the initial study phase (with the superimposed objects). Recognition was relatively good for previously attended objects. By contrast, explicit recognition was dramatically lower for previously ignored objects and did not differ from the rate of false "old" responses to new items (Fig. 4c).

A second group underwent the same study phase and was then tested on an unexpected task in which the participants had to identify visual objects from fragmented pictures (Snodgrass and Feenan 1990), without requiring any explicit judgment of whether these were old or new images compared to those shown in the preceding study phase. Identification was significantly worse for new (i.e., needed more complete versions of the objects) than for old objects (Fig. 4d). Performance was better for old objects from the previously attended stream than for those from the ignored stream, but critically, all previously ignored objects yielded better identification relative to that of new objects. These behavioral tests therefore demonstrated complete amnesia for previously ignored items on direct explicit testing and reliable behavioral priming on indirect testing, indicating that memory traces were formed for these objects even in the absence of attention.

A third group participated in an event-related fMRI study that measured neural responses to previously attended or ignored objects, shown alone in the same or mirror-reversed orientation, intermixed with new items as before. Repetition-related decreases in fMRI responses (see Vuilleumier et al. 2002a) to previously attended objects repeated in the same orientation were found in right posterior fusiform, lateral occipital, and left inferior frontal cortex. More anterior fusiform regions showed repetition-decreases for all "old" objects, irrespective of attention and orientation (Fig. 4e), confirming that ignored stimuli had attained relatively high-levels of visual processing during the study. In addition, previously ignored objects produced fMRI response-increases in bilateral lingual gyri (V1) relative to both previously attended and new objects (Fig. 4f), suggesting a selective effect of prior attentional suppression for these objects on the subsequent response in the early visual cortex (see below).

Selective attention at exposure can thus produce several distinct, long-term effects on visual processing of stimuli that are repeated later. Previously attended objects led to neural response-suppression and previously ignored objects resulted in some response-enhancement, both effects arising in different brain areas. Enhancement of previously ignored items cannot be explained by nonspecific changes from inattention in the first study phase to attention during the test phase since other brain regions (e.g., fusiform cortex) showed repetition decreases for the very same items during the test phase, and entirely new items did not produce such effects. The repetition enhancements observed here for ignored objects might be related to negative priming effects such as those observed in some behavioral tasks. Typically, when a previously ignored object later becomes a target, reaction times to this "new" target are slower (Tipper 1985). Negative priming may arise particularly in situations when attention serves to exclude distractors in favor of the target (Tipper and Driver 1988) and when explicit

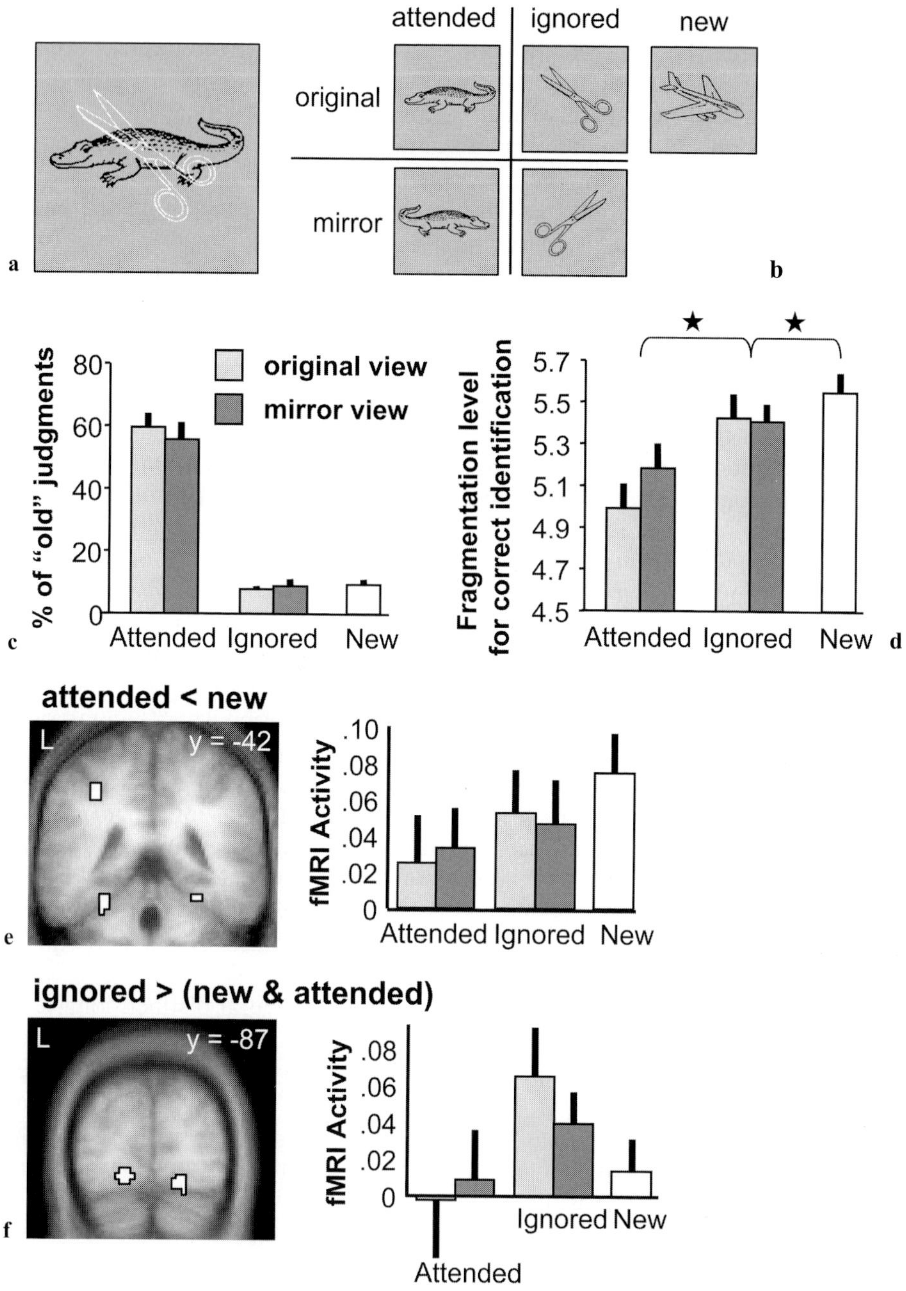
a
attended
ignored
new
original
mirror
b
% of "old" judgments
80
60
40
20
0
original view
mirror view
Attended Ignored New
c
Fragmentation level for correct identification
5.7
5.5
5.3
5.1
4.9
4.7
4.5
Attended Ignored New
d
attended < new
L
y = -42
fMRI Activity
.10
.08
.06
.04
.02
0
Attended Ignored New
e
ignored > (new & attended)
L
y = -87
fMRI Activity
.08
.06
.04
.02
0
Ignored New
Attended
f

FIG. 4. **a** In the first study phase, subjects saw a rapid visual stream of displays, each containing two overlapping shapes, one drawn in cyan, the other in magenta (here rendered in black and white). The task was to monitor only stimuli of a particular color (here the black ones). This study-phase was equal in all the behavioral and fMRI testing. **b** During the behavioral recognition task and during fMRI, black line drawings were presented one at a time, including previously attended and ignored objects, half of which were shown with the same view as in the first study part and half mirror reversed, randomly intermingled with new objects. **c** Percentage of "old" recognition judgments for each stimulus condition; the results show reliable explicit memory for previously attended items, and amnesia for previously ignored items. **d** Levels of picture fragmentation at which objects were correctly identified, showing significant priming for both previously attended and previously ignored items, as compared with new items. **e** Main effects of attended < new objects ("repetition suppression", irrespective of view change) found in bilateral anterior fusiform regions and the left intraparietal sulcus. Left: statistical SPM results overlapped on the mean normalized anatomical brain of participants; right: group-averaged fMRI response from the right anterior fusiform peak, showing repetition-decreases for both previously attended and ignored objects compared to new objects. **f** Activity in the bilateral lingual areas supportive of "repetition enhancement" for ignored verus new objects. Left: statistical SPM results superimposed on the mean anatomical brain of participants; right: group-averaged fMRI response from the left lingual peak, showing selective repetition-increases for previously ignored objects, but not for previously attended or new objects. Adapted from Vuilleumier et al. (2005)

awareness of the ignored stimulus is eliminated (Tipper 2001), as this was the case during our study phase. Moreover, negative priming can operate even for novel shapes (DeSchepper and Treisman 1996), which suggests that it might be involved in the early stages of shape processing, in accord with the lingual sites where fMRI repetition-enhancements were observed for ignored objects.

5 Conclusions

In this chapter, three fMRI studies were presented. Taken together, these studies provide converging evidence for significant functional plasticity, occurring at the first cortical stage of visual processing in the adult human brain. While the cellular mechanisms that underlie such long-term modulation of the BOLD signal in V1 remain largely unknown, recent theoretical models have proposed that perceptual learning might implicate local changes in excitatory and inhibitory influences within the visual cortex (e.g., Adini et al. 2002). Although our findings are mostly compatible with such models, these findings also attest to massive, top-down attentional influences from task-related requirements that impose significant constraints on long-term plasticity in V1. Therefore, models that would best fit our data need to incorporate both local influences within V1, and top-down influences from signal control and task expectations (e.g., Herzog and Fahle 1998).

Taken together, these data demonstrate an involvement of the early visual cortex in long-term effects of attentional selection and perceptual learning, thus challenging the traditional view of primary sensory cortices as hard-wired (see review by Fahle 2005). How permanent these neural changes are and to what extent they may also generalize to other stimuli or tasks, remain important questions for future research.

Acknowledgements. Supported by the Swiss National Science Foundation (grant #3100-AO-102133 to S.S.). Many thanks to all coauthors of the three studies reported here: R. J. Dolan, J. Driver, S. Duhoux, C. Frith, C. Hutton, P. Maquet, A. Maravita, and P. Vuilleumier. Grateful thanks to I. Rentschler, N. Osaka, and M. Osaka who organized an enlightening workshop in Kyoto (August 2004), where this work was presented.

References

Adini Y, Sagi D, Tsodyks M (2002) Context-enabled learning in the human visual system. Nature 415:790–793

Bahcall DO, Kowler E (1999) Attentional interference at small spatial separations. Vision Res 71–86

Brefczynski JA, DeYoe EA (1999) A physiological correlate of the "spotlight" of visual attention. Nat Neurosci 2:370–374

Chan HS, Courtney AJ (1998) Stimulus size scaling and foveal load as determinants of peripheral target detection. Ergonomics 1433–1452

Crist RE, Li W, Gilbert CD (2001) Learning to see: experience and attention in primary visual cortex. Nat Neurosci 4:519–525

DeSchepper B, Treisman A (1996) Visual memory for novel shapes: implicit coding without attention. J Exp Psychol Learn Mem Cogn 22:27–47

DeYoe EA, Carman GJ, Bandettini P, Glickman S, Wieser J, Cox R, Miller D, Neitz J (1996) Mapping striate and extrastriate visual areas in human cerebral cortex. Proc Natl Acad Sci U S A 93:2382–2386

Driver J, Frackowiak RS (2001) Neurobiological measures of human selective attention. Neuropsychologia 39:1257–1262

Engel SA, Glover GH, Wandell BA (1997) Retinotopic organization in human visual cortex and the spatial precision of functional MRI. Cereb Cortex 7:181–192

Fahle M (2005) Perceptual learning: specificity versus generalization. Curr Opin Neurobiol 15:154–160

Furmanski CS, Schluppeck D, Engel SA (2004) Learning strengthens the response of primary visual cortex to simple patterns. Curr Biol 14:573–578

Gandhi SP, Heeger DJ, Boynton GM (1999) Spatial attention affects brain activity in human primary visual cortex. Proc Natl Acad Sci U S A 96:3314–3319

Gilbert CD, Sigman M, Crist RE (2001) The neural basis of perceptual learning. Neuron 31:681–697

Grill-Spector K, Kushnir T, Edelman S, Avidan G, Itzchak Y, Malach R (1999) Differential processing of objects under various viewing conditions in the human lateral occipital complex. Neuron 24:187–203

Grossberg S, Williamson JR (2001) A neural model of how horizontal and interlaminar connections of visual cortex develop into adult circuits that carry out perceptual grouping and learning. Cereb Cortex 11:37–58

Herzog MH, Fahle M (1998) Modeling perceptual learning: difficulties and how they can be overcome. Biol Cybern 78:107–117

Holmes DL, Cohen KM, Haith MM, Morrison FJ (1977) Peripheral visual processing. Percept Psychophys 571–577

Hupe JM, James AC, Girard P, Bullier J (2001) Response modulations by static texture surround in area V1 of the macaque monkey do not depend on feedback connections from V2. J Neurophysiol 85:146–163

Ikeda M, Takeuchi T (1975) Influence of foveal load on the functional visual field. Percept Psychophys 255–260

Karni A, Sagi D (1991) Where practice makes perfect in texture discrimination: evidence for primary visual cortex plasticity. Proc Natl Acad Sci U S A 88:4966–4970

Kastner S, Ungerleider LG (2000) Mechanisms of visual attention in the human cortex. Annu Rev Neurosci 23:315–341

Kastner S, Pinsk MA, De Weerd P, Desimone R, Ungerleider LG (1999) Increased activity in human visual cortex during directed attention in the absence of visual stimulation. Neuron 22:751–761

Lavie N (2005) Distracted and confused?: selective attention under load. Trends Cogn Sci 9:75–82

Lavie N, Tsal Y (1994) Perceptual load as a major determinant of the locus of selection in visual attention. Percept Psychophys 56:183–197

Logothetis NK, Pauls J, Augath M, Trinath T, Oeltermann A (2001) Neurophysiological investigation of the basis of the fMRI signal. Nature 412:150–157

Merikle P, Smilek D, Eastwood JD (2001) Perception without awareness: perspectives from cognitive psychology. Cognition 79:115–134

O'Connor DH, Fukui MM, Pinsk MA, Kastner S (2002) Attention modulates responses in the human lateral geniculate nucleus. Nat Neurosci 5:1203–1209

O'Craven KM, Downing PE, Kanwisher N (1999) fMRI evidence for objects as the units of attentional selection. Nature 401:584–587

Parkin AJ, Reid TK, Russo R (1990) On the differential nature of implicit and explicit memory. Mem Cognit 18:507–514

Pinsk MA, Doniger GM, Kastner S (2004) A push-pull mechanism of selective attention in human extrastriate cortex. J Neurophysiol 98:622–629

Plainis S, Murray IJ, Chauhan K (2001) Raised visual detection thresholds depend on the level of complexity of cognitive foveal loading. Perception 1203–1212

Rees G, Frith CD, Lavie N (1997) Modulating irrelevant motion perception by varying attentional load in an unrelated task. Science 278:1616–1619

Rees G, Russell C, Frith CD, Driver J (1999) Inattentional blindness versus inattentional amnesia for fixated but ignored words. Science 286:2504–2507

Rock I, Guttman D (1981) The effect of inattention on form perception. Hum Percept Perform 275–285

Schwartz S, Maquet P, Frith C (2002) Neural correlates of perceptual learning: a functional MRI study of visual texture discrimination. Proc Natl Acad Sci USA 99:17137–17142

Schwartz S, Vuilleumier P, Hutton C, Maravita A, Dolan RJ, Driver J (2004) Attentional load and sensory competition in human vision: modulation of fMRI responses by load at fixation during task-irrelevant stimulation in the peripheral visual field. Cereb Cortex 15:770–786

Sereno MI, Dale AM, Reppas JB, Kwong KK, Belliveau JW, Brady TJ, Rosen BR, Tootell RB (1995) Borders of multiple visual areas in humans revealed by functional magnetic resonance imaging. Science 268:889–893
Smith AT, Singh KD, Greenlee MW (2000) Attentional suppression of activity in the human visual cortex. Neuroreport 11:271–277
Snodgrass JG, Feenan K (1990) Priming effects in picture fragment completion: support for the perceptual closure hypothesis. J Exp Psychol Gen 119:276–296
Somers DC, Dale AM, Seiffert AE, Tootell RB (1999) Functional MRI reveals spatially specific attentional modulation in human primary visual cortex. Proc Natl Acad Sci USA 96:1663–1668
Szymanski KF, MacLeod CM (1996) Manipulation of attention at study affects an explicit but not an implicit test of memory. Conscious Cogn 5:165–175
Tipper SP (1985) The negative priming effect: inhibitory priming by ignored objects. Q J Exp Psychol A 37:571–590
Tipper SP (2001) Does negative priming reflect inhibitory mechanisms? A review and integration of conflicting views. Q J Exp Psychol A 54:321–343
Tipper SP, Driver J (1988) Negative priming between pictures and words in a selective attention task: evidence for semantic processing of ignored stimuli. Mem Cognit 16:64–70
Tootell RB, Hadjikhani N, Hall EK, Marrett S, Vanduffel W, Vaughan JT, Dale AM (1998) The retinotopy of visual spatial attention. Neuron 21:1409–1422
Tsodyks M, Gilbert C (2004) Neural networks and perceptual learning. Nature 431:775–781
Vuilleumier P, Schwartz S, Husain M, Clarke K, Driver J (2001) Implicit processing and learning of visual stimuli in parietal extinction and neglect. Cortex 37:741–744
Vuilleumier P, Henson RN, Driver J, Dolan RJ (2002a) Multiple levels of visual object constancy revealed by event-related fMRI of repetition priming. Nat Neurosci 5:491–499
Vuilleumier P, Schwartz S, Clarke K, Husain M, Driver J (2002b) Testing memory for unseen visual stimuli in patients with extinction and spatial neglect. J Cogn Neurosci 14:875–886
Vuilleumier P, Schwartz S, Duhoux S, Dolan RJ, Driver J (2005) Selective attention modulates neural substrates of repetition priming and "implicit" visual memory: suppressions and enhancements revealed by fMRI. J Cogn Neurosci 17:1245–1260
Walker MP, Stickgold R, Jolesz FA, Yoo SS (2005) The functional anatomy of sleep-dependent visual skill learning. Cereb Cortex 15:1666–1675
Williams LJ (1984) Information processing in near peripheral vision. J Gen Psychol 201–207
Yantis S, Serences JT (2003) Cortical mechanisms of space-based and object-based attentional control. Curr Opin Neurobiol 13:187–193

3
Pattern Recognition in Direct and Indirect View

Hans Strasburger[1] and Ingo Rentschler[2]

1 Introduction

More than a century ago, it was shown that there is an acuity deficit in peripheral vision that can be compensated for by increasing stimulus size (Aubert and Foerster 1857; Wertheim 1894). The corresponding size-scaling approach, or cortical magnification concept, has accounted for much of the eccentricity variation in grating contrast sensitivity (Koenderink et al. 1978; Rovamo and Virsu 1979) and various other measures of acuity (e.g., Levi et al. 1985; Virsu et al. 1987). Yet this cannot be the whole truth since size-scaling fails to establish positional invariance for a wide range of visual tasks, like numerosity judgments (Parth and Rentschler 1984), discrimination of phase-modulated (Harvey et al. 1985) and mirror-symmetric images (Rentschler and Treutwein 1985), face recognition (Hübner et al. 1985), and recognition of numeric characters (Strasburger and Rentschler 1996); (Strasburger et al. 1991).

To explain this discrepancy, we previously suggested that peripheral vision ignores pattern structure independently of scale but detects image energy in much the same way as foveal vision does (Rentschler and Treutwein 1985; Rentschler 1985). Similarly, our previous study (1996) proposed that peripheral vision fails to integrate pattern features. Such explanations of functional inhomogeneity across the visual field remain somewhat vague as long as there is little known about the corresponding neural representation of patterns. To address that issue, we review two recent studies of pattern recognition in direct and indirect view, which used classification paradigms corresponding to two meanings of the term pattern recognition (cf. Watanabe 1985, Chap. 1): Strasburger (2005) elaborated on the recognition of numeric characters, i.e., the identification of patterns as members of already known classes. Jüttner and Rentschler (2000) investigated

[1]Generation Research Program, University of München, Bad Tölz, Germany, and Department of Medical Psychology, University of Göttingen, Waldweg 37, D-37073 Göttingen, Germany
[2]Institute of Medical Psychology, University of Munich, Goethestraße 31, D-80336 München, Germany

how observers learn to assign unfamiliar grey-level patterns to previously unknown classes.

2 Crowding Effect in Indirect View

A conspicuous limitation of pattern recognition on indirect view is known as the crowding effect, where performance is impaired for test patterns that occur in the presence of neighbouring patterns (Strasburger et al. 1991). The effect is small in foveal vision (Flom et al. 1963) but dramatically reduces recognition performance in extrafoveal vision (Bouma 1970). In amblyopia–a loss of visual function due to disuse in childhood–the effect is strong in the fovea as well (Stuart and Burian 1962). Crowding changes during visual development but shows a slower time course than that for acuity (Atkinson et al. 1986) and plays an important, if not fully understood, role in dyslexia (Geiger and Lettvin 1986). Figure 1 provides a simple demonstration of the effect.

The strong influence of retinal eccentricity on the crowding effect can be explained at least partly as an effect of spatial attention (Strasburger et al. 1991; He et al. 1996). This has been demonstrated by Strasburger et al. using a technique introduced by Averbach and Coriell (1961) who found a bar pointing towards the target letter, but not a circle around it, effective in directing the attention of observers to targets within letter strings. Thus, both spatial attention and lateral masking have been demonstrated. In addition, Strasburger and co-authors performed an error analysis similar to that by Eriksen and Rohrbaugh (1970), for separating sensory and attentional influences on lateral masking. Strasburger et al. succeed in showing that localization errors, i.e., the inadvertent reporting a flanker rather than the target, and failure to recognize the target character in the middle were equally frequent in many cases. They interpreted this result as a consequence of pattern recognition in the absence of positional information or the ability to precisely focus attention.

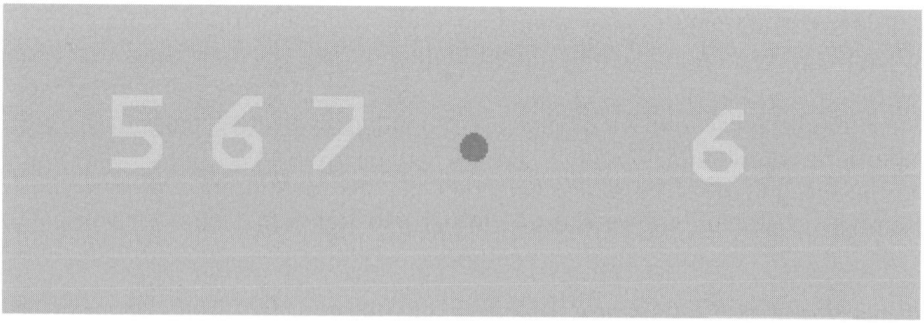

FIG. 1. Crowding effect. The two representations of the digit "6" are shown at the same contrast and distance from the fixation target. Yet, when vision is fixated on the dot, the "6" on the right is easily recognized, whereas the same "6" on the left is not

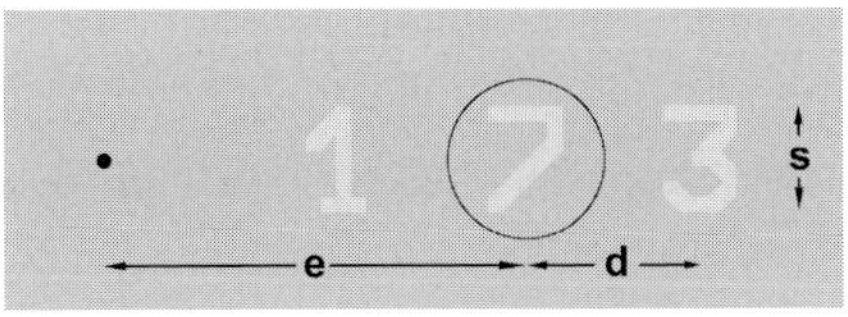

FIG. 2. Stimulus layout in the flanked and cued crowding condition. Letter size *(s)* is specified as letter height in degrees of visual angle; flanking distances *(d)* are measured from the respective character centres. *e*: eccentricity

Strasburger (2005) confirmed and extended these findings using three different recognition paradigms (Fig. 2). A standard crowding condition similar to that in Strasburger et al. (1991) was compared to a cued condition, which used a circle at the target position appearing just before the target, and a "content-only" condition, where positional information was separated from (semantic) pattern content. Contrast thresholds for the recognition of numeric characters (digits 0–9) were measured using an adaptive algorithm (Harvey 1997). Characters were presented in white on a grey background (50 cd/m^2 luminance) for 100 ms, either in isolation (baseline condition) or laterally flanked by two additional digits. Twenty observers of both genders (aged 20–30 years) were tested under identical conditions. In each crowding condition, three digits (target and flankers) of the same size and contrast were used. Subjects were instructed to report the middle digit, and the dependent variable was the contrast threshold for recognizing the letter. In the flanked condition, the target was surrounded by neighbouring digits. In the cued condition, a black circle was additionally exposed at the target location with an onset of 150 ms before the target. The circle was switched off at target onset.

The content-only condition was established by modifying the threshold criterion of the standard condition. Thresholds were determined by accepting as correct not only responses that identified the middle target but also responses that identified one of the flankers. Thus, subjects reflected the ability to recognize patterns independently of their location with sustained attention focused on the middle target. Taken together, there were two variations relative to the standard flanked condition (1): one, where spatial attention was modulated by a positional cue (2) and one, which separated target location and target content (3).

The magnitude of the crowding effect depends on stimulus size, character separation, contrast, and retinal eccentricity (Bouma 1970; Strasburger et al. 1991; Pelli et al. 2004). Three middle-character eccentricities, namely 1°, 2°, and 4°, were used with (scaled) stimulus sizes of 0.3°, 0.4°, and 0.6°, respectively. The size of the ring cues was scaled to 0.44°, 0.59°, and 0.88° in diameter.

Figure 3 shows the mean recognition thresholds over flanker distances under conditions (1)–(3). Thresholds for the single-digit are indicated by a horizontal line, together with the average standard error. As expected, of all three eccentricity conditions, (1) yields the highest thresholds. Crowding is absent at sufficiently large flanker distances, as seen in the top and middle graph of Figure 3, and

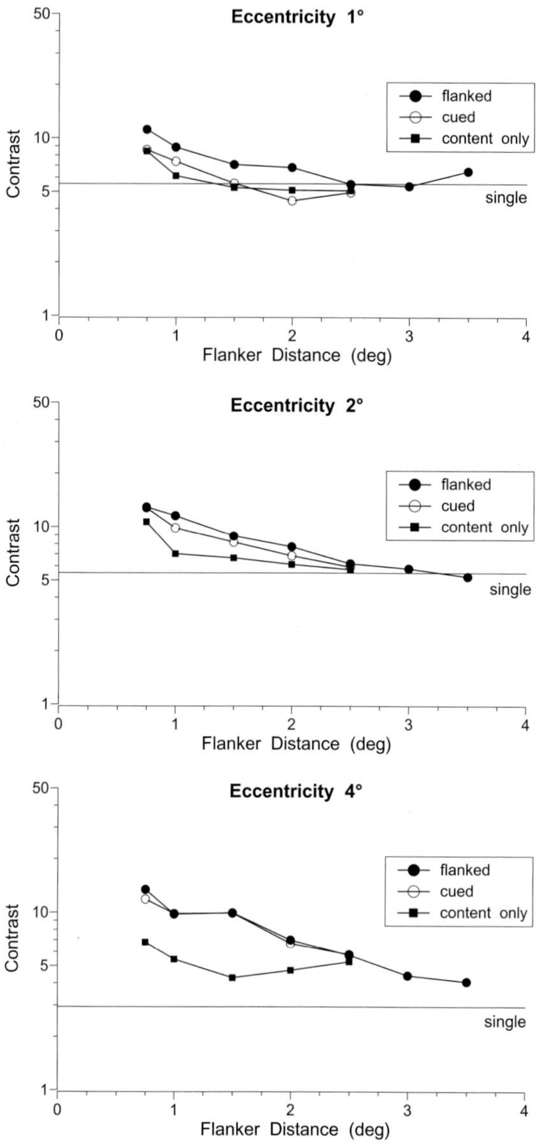

FIG. 3. Recognition contrast thresholds for the three crowding conditions as a function of flanker distance, at three eccentricities on the horizontal meridian (top to bottom graph 1°, 2°, and 4°, respectively). The thresholds for the single-character presentation are shown as thin horizontal lines; error bars on the corresponding data point show the mean for all data points in that sub-graph

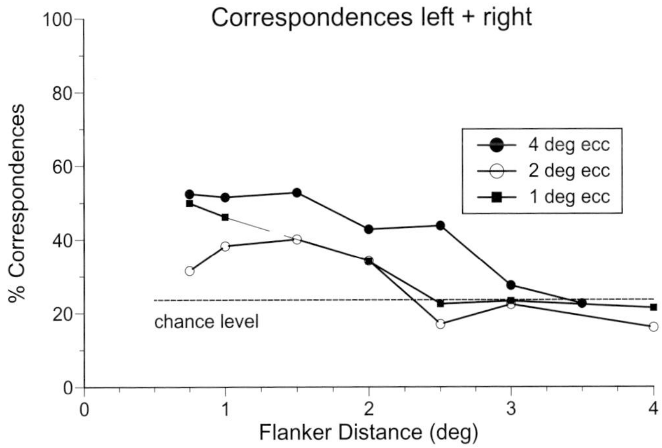

FIG. 4. Correspondences of the observers' incorrect responses with one of the flanking characters in the flanked condition, as a function of flanker distance. Chance level (23.6%) is indicated by the dashed line

gradually sets in at decreasing flanker distance. Contrast thresholds under condition (2) are below those of condition (1) at 1° and 2° eccentricity but still clearly above those under the single-digit condition. Thus, the ring cue was, at these eccentricities, partially effective in focusing attention on the middle character. Reasons for the cue not being effective at 4° could be a circle size that was too small, thus introducing some masking along with attention guidance (Averbach and Coriell 1961). Contrast thresholds are lowest (i.e. performance best) under the content-only condition (Fig. 3, filled squares). For eccentricities of 1° and 2°, thresholds are nearly equal to those corresponding to the single-digit condition (horizontal line). At 4° eccentricity, thresholds are elevated but still clearly below those of the standard flanking condition. Thus, when the position of a character within a letter string was ignored, its recognition under crowding conditions was almost as good as that when presented in isolation.

Figure 4 shows the results of error analysis. The dependent variable "correspondences" indicates how often a character, that was erroneously reported to be present at the target location, actually occurred as one of the flanking characters. Related chance performance (23.6%) is indicated by the dashed line in Figure 4. The difference between the proportion of correspondences and chance level can be attributed to localization errors, where observers correctly identified a pattern but missed its location. Such errors do not occur at large flanker distances and clearly increase with decreasing flanker distance. At their maximum, observed correspondences are as high as 52% (filled circles), thus demonstrating close to 30% recognitions at the wrong location (52%–23.6% chance). The remaining errors (100%–52% = 48%) can be attributed to a failure in recognizing pattern content. The comparison of Figures 3 and 4 further shows that flanker distances, below which crowding and mislocalization, respectively, take place, are about equal. Thus, localisation errors occur if and only if there is crowding.

The effects of crowding conditions on contrast thresholds and on correspondences were tested for statistical significance using two one-way analyses of covariance, with crowding condition as the factor (and linearized eccentricity as covariate). All effects were highly significant at the 1% level. Importantly, the correspondences are nearly equal (38.6% *vs* 39.1%) between the flanked (1) and the flanked-and-cued condition (2). Therefore, the cue is effective in improving recognition performance (as shown above under condition 2) but the improvement does not stem from moving attention away from the flanking characters.

To summarize, at flanker distances up to 2.5° (eccentricity ≤4°), the crowding effect is to a large part (up to 30%) explained by imprecise coding of the target character's position. Remaining errors (48%) can be attributed to insufficient coding of pattern content. A ring cue preceding the target enhances (content) recognition by sharpening transient spatial attention but leaves positional coding unaltered. Thus it appears that pattern identity and pattern location are separately encoded.

3 Attentional Spotlight and Feature Integration

As has previously been conjectured, the visual periphery seems to have a restricted ability to encode spatial relations between pattern components or integrate pattern features (Rentschler and Treutwein 1985; Strasburger and Rentschler 1996). Similarly, Pelli et al. (2004) characterized crowding as a process of impaired feature integration occurring in the visual periphery, in distinction to lateral masking from impaired feature detection occurring anywhere in the visual field. Strasburger (2005) proposed that the range of feature integration is related to spatial attention and might reflect the spread of attentional spotlight. Distinguishing sustained and transient visual attention (Nakayama and Mackeben 1989; Mackeben 1999), the standard crowding task involves *sustained* attention since subjects were well aware in advance of where the stimulus would appear. The role of the ring cue in that framework was to enhance content coding by increasing *transient* attention, leaving position coding unaffected.

How does the concept of attention mediating feature integration fit with neurophysiological findings? Flom et al. (1963) have shown that lateral interactions do also occur when target and flankers are presented to one eye and the other eye, respectively (dichoptic viewing conditions). Interactions therefore occur at the cortical stage. Results of dichoptic masking in the fovea and in the periphery support this view (Tripathy and Levi 1994). Strasburger (2005) elaborated on that within a concept of attention involving the spatially selective control of bottom-up activation through top-down connections. Selectivity was assumed to be mediated by retinotopically organized brain structures (cf. LaBerge 1995; Vidyasagar 2001). The gating itself could occur in early cortical areas or even in the lateral geniculate nucleus. The latter is commonly thought to subserve a gating function in the retino-cortical pathway. Indeed, Vidyasagar has shown attentional modulation in single-cell studies as early as in V1 (see also the chapter by S. Schwartz, this volume).

These observations suggest that, mediated through the pulvinar and V1, brain regions involved in attention selectively control retinotopically organized bottom-up activation. Owing to the function of a winner-take-all network (perhaps subserving Gestalt closure and related to object-based attention), the dominant stimulus representation might be selectively relayed to cortical areas performing visual feature integration like the inferotemporal cortex (ITC, see Tanaka 1996). Feature integration could occur in an unintended region of the visual field if the information encoded in the neural map is imprecise in location or spatial extent. In such cases, the perceived pattern would not coincide with the target. The ring cue, however, would seem to pre-activate the corresponding (retinotopic) location in the map without affecting other locations.

4 Category Learning *vs* Discrimination Learning

To explore pattern encoding in direct and indirect view, Jüttner and Rentschler (1996, 2000) used a paradigm of supervised learning, where unfamiliar grey-level patterns ("compound Gabor signals") are assigned to a given number of pattern classes. The luminance profiles of stimuli were varied through the modulation of phase relationships and, to some extent, amplitudes between spatial frequency components. Resulting classification tasks therefore largely involved the distinction of pattern structure.

Two types of classification tasks were compared, each involving a learning set of 15 patterns. Learning patterns were to be assigned to three classes having fixed mean pattern vectors for identical image energy (Fig. 5). Set A, with a large variance in signals within each class and relatively small variance of signals between classes, presented participants with a difficult task. Set B, with a small variance within classes and large variance between classes, presented subjects with an easy task (Fig. 5a, left). Discrimination tasks involved the same stimulus sets used in three consecutive experiments, each requiring observers to assign sub sets of 10 learning signals to one of two pattern classes (Fig. 5b). Discrimination tasks thus conformed to the Delayed-Matching-to-Sample paradigm of behavioural research (see Miller and Desimone 1994). Three viewing conditions were employed: pattern exposure at the locus of fixation (central) and fixation 3° to the left and 3° to the right of the pattern centre, respectively (left and right). Pattern size was scaled according to cortical magnification (Rovamo and Virsu 1979). Learning performance was characterized using the number of learning units to criterion and a computational model providing mappings of internalized pattern representations onto their physical counterparts (probabilistic virtual prototypes, PVP; Rentschler et al. 1994).

PVP solutions for discrimination learning are obtained by making use of the fact that "dipole configurations" of pairs of pattern representations can be combined as in vector addition (see Jüttner and Rentschler 1996, Appendix I). Thus, it is demonstrated that such solutions for discrimination learning veridically

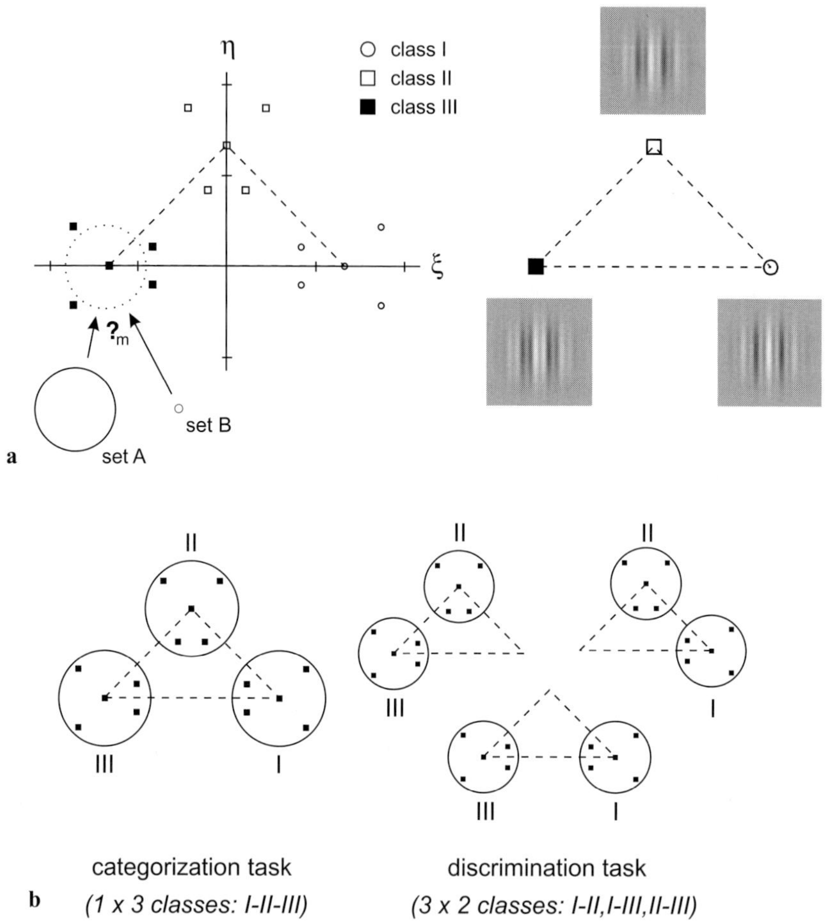

FIG. 5. **a** Pattern stimuli for discrimination and category learning. Stimulus sets consisted of 15 compound Gabor signals defined by a cosine waveform and its third harmonic, both modulated by an isotropic Gaussian aperture. The third harmonic was varied in amplitude b and phase φ. The physical signal representation used the features of evenness, $\eta = b \cos \varphi$, and oddness, $\xi = b \sin \varphi$. Pattern classes consisted of three clusters of five samples each. Scale: 1 unit = $15\,cd\,m^{-2}$. Mean pattern luminance $70\,cd\,m^{-2}$. Right: Images corresponding to the mean vectors of pattern classes. **b** Category learning (left) involved three pattern classes simultaneously. Discrimination learning (right) successively involved pairs of pattern classes (reproduced with permission from Jüttner and Rentschler 2000)

represent the physical signal configurations in both direct and indirect view for both difficult and easy tasks (Fig. 6, first rows for sets A and B). Similar results are obtained for category learning with the easy stimulus set B (Fig. 6, second row for set B). For the difficult set A, however, quasi-congruence of physical signal configurations and reconstructed pattern representations is only obtained for stimulus exposure at the locus of fixation (Fig. 6, set A, second row, centre).

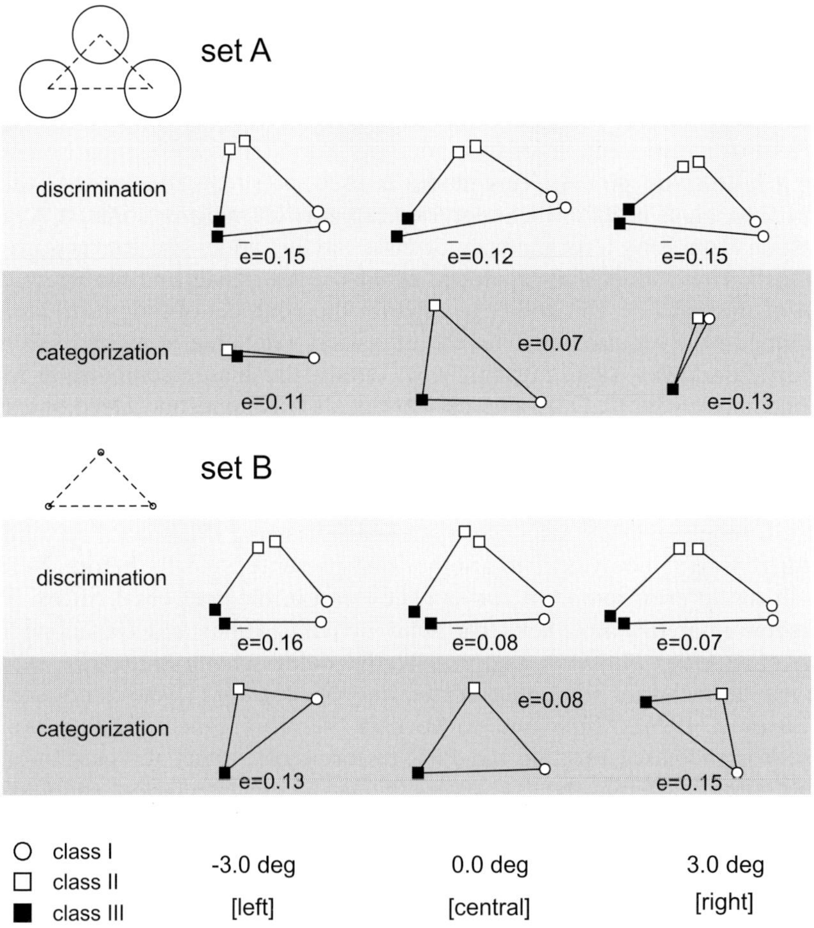

FIG. 6. Pattern representations generated through discrimination and category learning in direct and indirect view. Data obtained by re-projecting "virtual" class prototypes from behavioural classification data in physical feature space by means of probabilistic Bayesian classifiers (reproduced with permission from Jüttner and Rentschler 2000)

For off-axis stimulation, pattern representations degenerate to linear configurations, lacking extent in the second dimension (Fig. 6, set A, second row, left and right). Similarly, learning duration was massively prolonged (about 8-fold) for category learning with the difficult stimulus set A under conditions of off-axis observation only.

These results falsify our original hypothesis according to which relational pattern encoding is impossible in indirect view. Instead, they indicate that structure-based discrimination and easy categorization tasks can be performed in indirect view provided size-scaling is applied. Yet there is an inability to perform difficult pattern classification tasks by indirect view that occurs even when sampling characteristics are accounted for by size-scaling.

5 Object Selective Attention in Direct and Indirect View

Concerning the difference between discrimination learning and category learning in direct and indirect view, it is noteworthy that for discrimination learning it is sufficient to construct in short-term memory a model of the comparison signal using a bottom-up process. This model is used as a template against which a comparison signal is matched. Supervised category learning, by contrast, requires activation from long-term memory of class models under the instructions of a "teacher". These models are matched against input signals and modified in the event of mismatches. The ability to activate the contents of long-term memory according to the requirements of current tasks is a defining property of working memory (Baddeley 1986; Fuster 2003). Within the biased-competition model (see the chapter by G. Deco and co-workers, this volume, and Deco and Rolls 2004), representations of pattern categorization in working memory can be considered templates for object selective attention (Desimone and Duncan 1995).

In the monkey, DMS tasks were studied by having the animal recognize a stimulus as being equivalent to another one presented shortly before. Neurons both in the inferior temporal cortex (ITC) and in the prefrontal cortex (PFC) may show related sample-selective delay activity (Miller and Desimone 1994; Miller et al. 1996). However, sample-selective delay activity in the PFC survives intervening irrelevant stimulus pairings, whereas in the ITC, this is not the case (Miller et al. 1996). Miller and co-workers therefore concluded that working memory is mediated through the PFC in terms of explicit representations of sample stimuli, whereas the ITC allows the automatic detection of stimulus repetitions only.

It is tempting, therefore, to speculate that, in distinction from discrimination learning, human category learning relies on pattern representations in working memory as have been found in the PFC of the monkey by Miller and colleagues. Our findings would then imply that memories from pattern stimulation in the peripheral visual field are not only spatially under-sampled due to cortical magnification but can also be activated in working memory to a restricted extent only. Thus, we suggest that the restricted ability to classify complex patterns in indirect view reflects a restricted capacity of object-selective attention.

6 Structured Pattern Representations in Direct and Indirect View

The proposal of pattern classification involving the representation of class models in working memory warrants further consideration. Traditional approaches to pattern recognition are based on the notion that members of a given class share certain features or feature vectors. Such descriptions allow the classification of simple isolated patterns but problems arise as pattern complexity increases and/

or patterns are embedded in scenes. Feature vectors are then found to be inadequate for encoding the variability of class samples. One reason for this is that patterns of different classes may share common feature vectors yet be structurally different (Bischof and Caelli 1997). This difficulty led to the development of structural or syntactic pattern recognition that underlies more recent approaches to object recognition within the domain of machine intelligence (see Caelli and Bischof 1997). Strategies of learning structural pattern representations based on part attributes (unary features) and part relations (binary features) have been developed in that context. Moreover, such strategies have been employed successfully for the analysis of category learning by humans (see the chapter by M. Jüttner, this volume).

With regard to pattern classification in direct and indirect view, it is important to note that there are several types of part-based recognition strategies (see Caelli and Bischof 1997). Such strategies may use "attribute-indexed" representations only, thus ignoring the associations between features and pattern parts. For instance, two patterns may be distinguished by a difference in the distributions of distances between pattern parts. In case of mirror-image signals, however, these distributions would be identical as such patterns are characterized by the same sets of unary features and (undirected) binary features. The classification of mirror-image patterns therefore requires "part-indexed" representations providing explicit associations between relational attributes and the pattern parts these refer to. Part-indexed representations for visual pattern recognition may be implemented using the attribute of "position" relative to an allocentric or scene-based frame of reference (Rentschler and Jüttner 2007). In general, part-indexed representations allow for more powerful but computationally more expensive strategies of structural pattern processing (Caelli and Bischof 1997).

It might be hypothesized, therefore, that pattern recognition in indirect view relies on attribute-indexed representations only, whereas attribute-indexed as well as part-indexed representations are available in direct view. Consistent with this proposal would be an inability to distinguish mirror-image patterns in extrafoveal vision (Rentschler and Treutwein 1985; Rentschler 1985; Saarinen 1987). It is impossible, however, to decide whether such a functional restriction of recognition on indirect view could be attributed to a limitation with regard to the access to working memory or origination at earlier stages of visual processing.

7 Conclusions

The size-scaling concept fails to account for the functional inferiority of peripheral vision in a wide range of pattern recognition tasks. We have hypothesized in the past that this can be explained, additional to a coarser grain, by an inability to properly integrate pattern features or encode structure. Here we have reviewed more recent findings demonstrating the possibility of recruiting or learning structured representations for pattern recognition in direct and indirect view. Yet we delimited the following shortcomings of pattern recognition on sideways viewing:

(1) The recognition of numerical characters in indirect view depends on whether digits occur in isolation or in combination with flanking characters (crowding effect). The interference of distractors and spatial cueing with the recognition of target characters indicates separate neural encoding of semantic pattern content and pattern position within certain spatial arrays, possibly based on a limitation of spatial selective attention.

(2) Peripheral vision not only fails in distinguishing mirror-symmetric patterns but also in solving difficult tasks of structure-based pattern classification. The latter type of functional restriction can be attributed to a limited access to working memory or, in other terms, of object selective attention. It is not clear, however, whether the difficulty with mirror images is a consequence of this limitation or originates at an earlier level of visual processing.

These findings are consistent with the view that objects are represented in the brain at several levels from the sensory to the semantic (cf. Fuster 2003), with different mechanisms of attention operating at each of these levels (cf. Desimone and Duncan 1995).

Acknowledgements. We are indebted to Sophie Schwartz, Gustavo Deco, Naoyuki Osaka, and Keiji Tanaka for helpful discussions of our work.

References

Atkinson J, Pimm-Smith E, Evans C, Harding G, Braddick O (1986) Visual crowding in young children. Doc Ophthalmol Proc 45:201–213

Aubert H, Foerster CFR (1857) Beiträge zur Kenntnis des indirecten Sehens. (I). Untersuchungen über den Raumsinn der Retina. Arch Ophthalmol 3:1–37

Averbach E, Coriell AS (1961) Short-term memory in vision. Bell System Tech J 40:309–328

Baddeley A (1986) Working memory. Clarendon Press, Oxford

Bischof WF, Caelli T (1997) Scene understanding by rule evaluation. IEEE Trans Pattern Anal Machine Intell (PAMI) 19:1284–1288

Bouma H (1970) Interaction effects in parafoveal letter recognition. Nature 226:177–178

Caelli T, Bischof WF (1997) Machine learning and image interpretation. Plenum Press, New York

Deco G, Rolls ET (2004) A neurodynamical cortical model of visual attention and invariant object recognition. Vision Res 44:621–642

Desimone R, Duncan J (1995) Neural mechanisms of visual attention. Annu Rev Neurosci 18:193–222

Eriksen CW, Rohrbaugh JW (1970) Some factors determining efficiency of selective attention. Am J Psychol 83:330–343

Flom MC, Weymouth FW, Kahnemann D (1963) Visual resolution and contour interaction. J Opt Soc Am 53:1026–1032
Fuster JM (2003) Cortex and mind. Oxford University Press, Oxford
Geiger G, Lettvin JY (1986) Enhancing the perception of form in peripheral vision. Perception 15:119–130
Harvey LO, Jr. (1997) Efficient estimation of sensory thresholds with ML-PEST. Spat Vis 11:121–128
Harvey LO, Jr., Rentschler I, Weiss C (1985) Sensitivity to phase distortion in central and peripheral vision. Percept Psychophys 38:392–396
He S, Cavanagh P, Intriligator J (1996) Attentional resolution and the locus of visual awareness. Nature 383:334–337
Hübner M, Rentschler I, Encke W (1985) Hidden-face recognition: comparing foveal and extrafoveal performance. Hum Neurobiol 4:1–7
Jüttner M, Rentschler I (1996) Reduced perceptual dimensionality in extrafoveal vision. Vision Res 36:1007–1022
Jüttner M, Rentschler I (2000) Scale-invariant superiority of foveal vision in perceptual categorization. Eur J Neurosci 12:353–359
Koenderink JJ, Bouman MA, Bueno de Mesquita AE, Slappendel S (1978) Perimetry of contrast detection thresholds of moving spatial sine wave patterns. I. The near peripheral visual field (eccentricity 0°–8°). J Opt Soc Am 68:845–849
LaBerge D (1995) Computational and anatomical models of selective attention in object identification. In: Gazzaniga MS (Ed) The cognitive neurosciences. MIT Press, Cambridge MA, pp 649–663
Levi DM, Klein SA, Aitsebaomo AP (1985) Vernier acuity, crowding and cortical magnification. Vision Res 25:963–977
Mackeben M (1999) Sustained focal attention and peripheral letter recognition. Spat Vis 12:51–72
Miller EK, Desimone R (1994) Parallel neuronal mechanisms for short-term memory. Science 263:520–522
Miller EK, Erickson CA, Desimone R (1996) Neural mechanisms of visual working memory in prefrontal cortex of the macaque. J Neurosci 16:5154–5167
Nakayama K, Mackeben M (1989) Sustained and transient components of focal visual attention. Vision Res 29:1631–1647
Parth P, Rentschler I (1984) Numerosity judgements in peripheral vision: limitations of the cortical magnification hypothesis. Behav Brain Res 11:241–248
Pelli DG, Palomares M, Majaj NJ (2004) Crowding is unlike ordinary masking: distinguishing feature integration from detection. J Vis 4:1136–1169
Rentschler I (1985) Symmetry-coded cells in the visual cortex? Nature 317:581–582
Rentschler I, Jüttner M (2007) Mirror-image relations in category learning. Vis Cognit 15:211–237
Rentschler I, Treutwein B (1985) Loss of spatial phase relationships in extrafoveal vision. Nature 313:308–310
Rentschler I, Jüttner M, Caelli T (1994) Probabilistic analysis of human supervised learning and classification. Vision Res 34:669–687
Rovamo J, Virsu V (1979) An estimation and application of the human cortical magnification factor. Exp Brain Res 37:495–510
Saarinen J (1987) Perception of positional relationships between line segments in eccentric vision. Perception 16:583–591

to image understanding and scene analysis (see e.g., Shapiro and Haralick 1981; Fan et al. 1989). These challenges inspired the development of a new generation of part-based recognition schemes (e.g., Jain and Hoffman 1988; Caelli and Dreier 1994; Rivlin et al. 1995; Bischof and Caelli 1997; Pearce and Caelli 1999). Such schemes can be classified as *generic part-based approaches* as they use the term "part" in a more flexible way. Parts may be defined either in the 2D image domain or based on the 3D range data that includes depth information. Furthermore, analysis into parts is seen as a recursive concept – applicable in the initial segregation of a scene into object(s) and background as well as in object classification and object identification. It is this versatility that makes generic part-based approaches so attractive for the cognitive modelling of human performance. The present paper focuses on one such approach, evidence-based systems (EBS), that allows the development of a process model for human category learning. Its usefulness is demonstrated by analysing behavioural data in experiments that address the effects of context and of mirror-image relations in pattern category learning – aspects that are difficult to assess by traditional psychometric categorization models. Finally, the principles and potential of generic part-based strategies, as exemplified by EBS, are related to current standard models of human object recognition.

2 Evidence-Based Pattern Classification and Category Learning

The EBS approach is based on the notion that complex patterns are best encoded in terms of parts and their relations (Jain and Hoffman 1988; Caelli and Dreier 1994). In an evidence-based classification system, a given pattern is first segmented into its component parts (Fig. 1). Each part is characterized by a vector of part-specific, or unary, attributes (e.g., size, luminance, area), and each pair of parts is described by a vector of part-relational, or binary, attributes (e.g., distance, angles, contrast). Within each attribute space, regions that function as activation regions for rules are defined. These regions result from the distribution of the attribute vectors of all objects used in the database during training. Once triggered by an attribute vector, the activation of a given rule provides a certain amount of evidence for the class membership of the input object. Evidence weights are implicitly estimated via the weights of a three-layer neural net. Here each input node corresponds to a rule, each output node to a class, and there is one hidden layer. The relative activity of an output node provides a measure of the accumulated class-specific evidence, which can be probabilistically interpreted and related to a classification frequency.

When used as a framework for cognitive modelling (see Jüttner et al. 1997, 2004), EBS describes category learning as the successive testing of working hypotheses. Each working hypothesis corresponds to the selection of a subset of

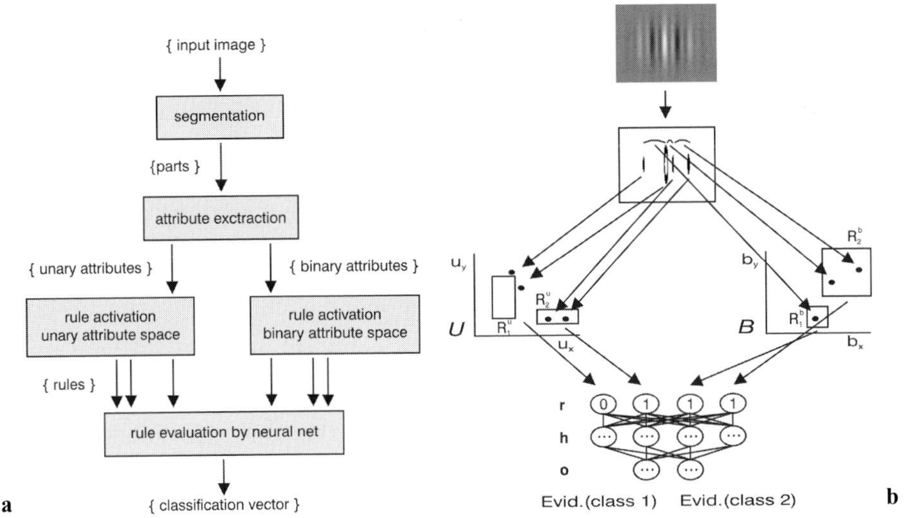

FIG. 1a,b. **a** Processing stages and **b** sample illustration of the representational levels involved in evidence-based classification. For convenience, only two unary and binary attributes, four rules and two classes are considered. See main text for further details. (From Jüttner et al. 1997, with permission)

attributes, which define a reference system for describing parts and their relations. Once chosen, the elaboration of such a working hypothesis will include the formation of rules and the tuning of evidence weights. Eventually, the elaboration process results in a successful categorization. Otherwise, the current working hypothesis is rejected and replaced by a different one.

Each subset of attended attributes may be regarded as a state within a search space of possible working hypotheses defined by the set of all combinations of unary and binary attributes. Learning speed is determined by the time required to find a solution within that search space and should be directly proportional to N_{FS}/N, where N_{FS} denotes the number of EBS solutions within the search space and N the total numbers of states. When used as a predictor for behavioural learning time, any variation of the categorization task that affects learning difficulty (i.e., the number of EBS solutions) should be reflected in the observed learning duration, which should vary according to $1/N_{FS}$. Therefore, the set of attributes evaluated for rule generation can be regarded as a "signature" of the underlying conceptual representation.

The two key features of EBS modeling, i.e., the analysis of learning dynamics and the reconstruction of categorical knowledge structures, will be demonstrated here in two behavioural experiments addressing the impact of context on category representations and the role of mirror-image relations in category learning.

3 EBS Applications I: The Impact of Context on Category Learning and Representation

Although our perception of the world is highly adaptive to contextual information, *context* has remained a relatively vague concept in vision research. Previous research has considered the role of context mainly with regard to visual selection and object recognition. With regard to the former, contextual information has been shown to modulate the deployment of spatial attention (Chun and Jiang 1998) and the statistical pattern of saccadic eye movements (Morris 1994; De Graef 1990). At a more cognitive level, many studies have demonstrated that identification is facilitated when an object is semantically consistent rather than inconsistent with the scene in which it appears (e.g., Palmer 1975; Biederman 1981), although the level of processing at which the contextual modulation of perception occurs has remained controversial (see e.g., Biederman 1972; Palmer 1975; Henderson et al. 1999).

The approaches described above all regard context as a determinant of how previously acquired knowledge guides the interpretation of sensory experience. In this study, we pursued a complementary perspective by showing that context also specifically affects *learning*, that is the acquisition of knowledge and the way in which such knowledge is mentally represented. For visual perception, such learning involves in particular the acquisition of object categories as the basis of object recognition (Rosch 1978).

In two classification-learning experiments, we explored the effect of complementary manipulations of context on the internal representation of pattern categories. Our experimental paradigm involved the classification of Compound Gabor patterns (Fig. 2), which ensured that the learning process was entirely under experimental control, as such stimuli are unfamiliar to naive observers.

In Experiment 1, we compared category learning and generalization with respect to two different class configurations (Fig. 2a): a 3-class configuration defined by three clusters of five patterns each (set 1), and a 4-class configuration composed of four clusters of three patterns (set 2). In the first part of the experiment, subjects were trained using a supervised-learning schedule to correctly classify all patterns of one of the two learning sets. Learning was partitioned into learning units and each learning unit consisted of two phases, a training phase and a test phase. During training, each pattern was presented three times in random order for 200 ms, followed by the corresponding class number displayed for 1 sec. During testing, each pattern was presented once in random order and was classified by the observer. Once the subjects had reached the learning criterion, they entered the second part of the experiment. Here the observers' ability to recognize contrast-inverted versions of the previously learned patterns was assessed. Each test pattern was presented and classified 30 times in random order. The timing parameters were the same as those used in the learning part of the experiment.

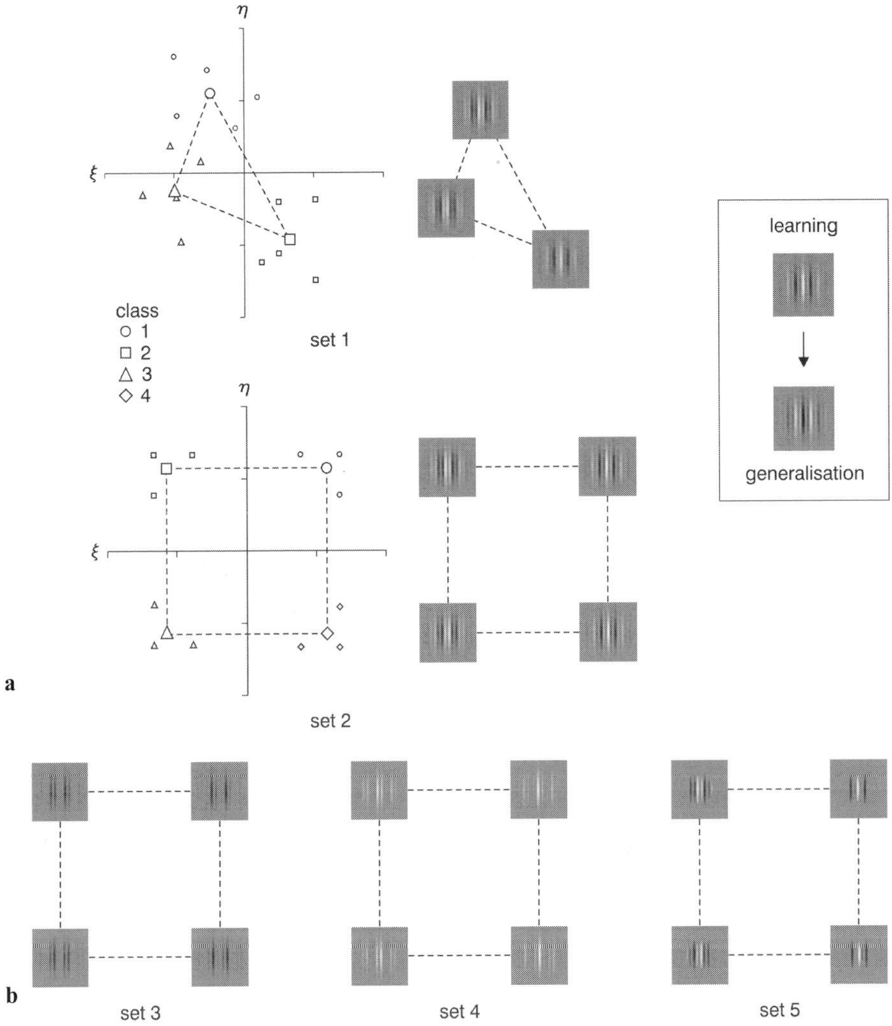

FIG. 2a,b. Experiments on context effects in category learning. **a** For Experiment 1, two sets of Compound Gabor patterns were generated in a two-dimensional evenness-oddness Fourier space (for details see Jüttner and Rentschler 1996). Scale: 1 unit = 20 cd/m^2. Set 1 contained three clusters of five samples, set 2 four clusters of three samples. Each signal cluster defined one class to be learned by the subject. The large symbols connected by dashed lines denote the class means or prototypes. For illustration, these class prototypes are depicted in their greylevel representation. **b** For Experiment 2, the 12 patterns of set 2 were degraded by replacing parts of the grey levels by mean luminance. Thus, in set 3, all bright parts of the image were removed, in set 3, all dark parts of the image, and in set 5, all intermediate values. In order to assess generalization, for each learning set, a corresponding set of test patterns was generated by inverting the contrast polarity of the patterns as indicated in the inset. (From Jüttner et al. 2004, with permission)

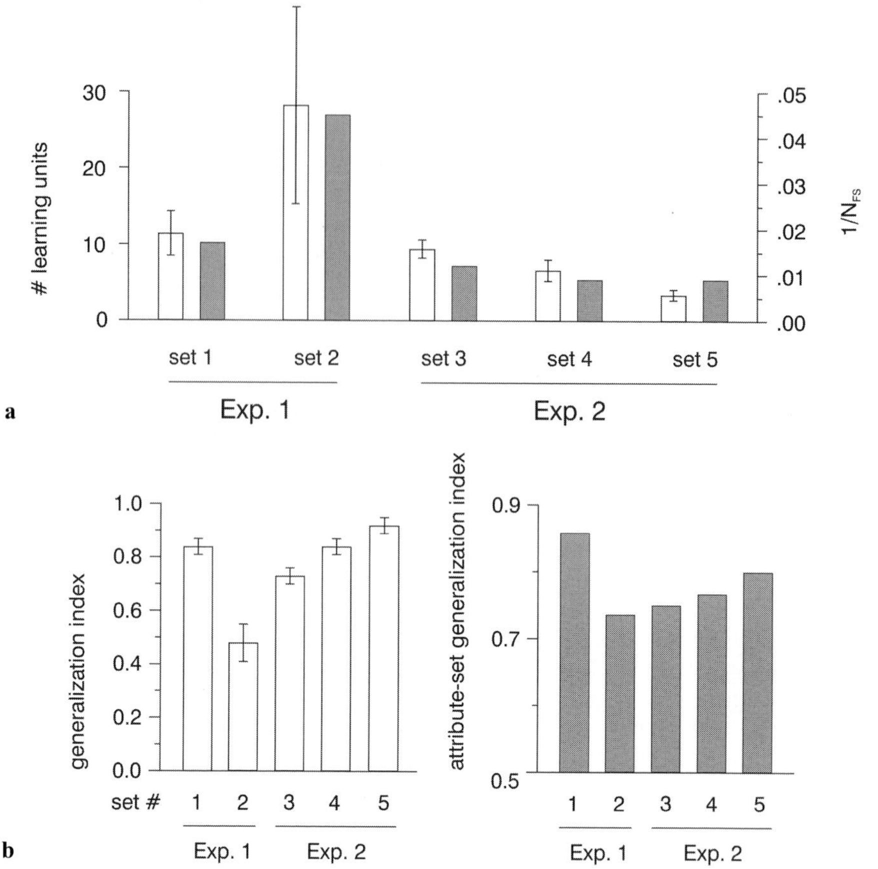

FIG. 3a,b. **a** EBS-simulated learning durations ($1/N_{FS}$, where N_{FS} is the number of EBS solutions in the learning space, *dark bars*) and observed group means *(bright bars)* for learning set 1–5 in Experiments 1 and 2. **b** Empirical generalization performance (*bright bars*, cf. Fig. 3) and EBS attribute-generalization indices *(dark bars)* for set 1–5 in Experiments 1 and 2

Two groups of five subjects with normal or corrected-to-normal vision participated. Figure 3a (bright bars) and Figure 3b (left) show learning duration and generalization performance for the two sets of learning patterns. The data demonstrated that the different learning contexts, as expressed by the two classes of configurations, had distinct effects on both performance indices. On average, the 3-class configuration was learned within 12 learning units, whereas the subjects needed about 28 learning units to learn the 4-class configuration. For generalization, the response rate for correctly classifying the contrast-inverted images was about 0.8 for set 1, and dropped to 0.45 for set 2.

In Experiment 1, we changed the context *locally*, i.e., for the individual pattern, by changing the configuration of the learning set. In contrast, Experiment 2 involved a *global* change of context, i.e., a manipulation of context for a complete class configuration. This was achieved by modifying the degree of stimulus accentuation. Based on the 4-class configuration (set 2) of Experiment 1, three further sets of learning patterns were generated by removing image parts via a thresholding operation (Fig. 2b). With these new sets of stimuli, we trained three groups of subjects to criterion and tested them for their ability to generalize to contrast inversion.

The accentuation brought about by thresholding yielded a drastic reduction of learning duration, from about 28 learning units for the original signals in set 2 to about 4 learning units for those of set 5 (bright bars in Fig. 3a). At the same time, generalization to contrast inversion improved, from about 0.45 in set 2 to about 0.95 (Fig. 3b left). Thus, the accentuated versions of the learning stimuli were much easier to learn and facilitated better generalization than the original images.

The variations of learning context introduced in Experiment 1 and Experiment 2 yielded distinct effects on both learning and generalization performance. To gain further insight into the nature of the underlying mental representations, we modeled human performance in terms of evidence-based pattern classification. For the simulations, we constrained the system parameters in a way that had proved optimal in previous work involving the same type of stimulus material (see Jüttner et al. 1997, 2004): The segmentation stage used a region-analysis technique that was based on partitioning the image according to connected grey-level regions yielding 3–5 parts per image. The rule-generation stage employed a K-means clustering procedure producing a set of 10–14 rules. The classifier was supplied with a reservoir of four unary attributes (position, luminance, aspect ration and size) and three binary attributes (distance, relative size, contrast). We then tested at which attribute combinations the training of the system converged, i.e., the system that successfully promoted the ability to distinguish between the classes.

The model predicted learning durations for the five sets of patterns are summarized by the dark bars of Figure 3a. A comparison with the behavioural data (bright bars) shows that for both experiments not only was the ranking order of the empirical learning durations preserved in the simulated values, but also the ratios of learning durations were well approximated by the latter. Thus, the model provides a unified account for context effects induced by very different experimental manipulations – the alteration of class configuration (Experiment 1) and variations of stimulus degradation (Experiment 2).

The analysis in terms of EBS also allows predictions concerning the difficulty of generalizing acquired class knowledge to contrast inversion. For this purpose, we computed an attribute-generalization index, defined as the ratio of the sum of contrast-independent attributes and the overall sum of all attributes appearing in the EBS solutions. As illustrated in Figure 3b (right), the ranking of this index mirrors the ranking in observed generalization performance. The proportion of contrast-invariant attributes therefore determines how well class concepts relying on these attributes may be generalized to contrast inversion.

4 EBS Applications II: Mirror-Image Relations in Category Learning

Visual patterns that are mirror-symmetric counterparts of each other are notoriously difficult to distinguish. Our ability to learn such distinctions to the point where they become almost trivial (for example, the distinction between the letters p and q, or "left shoe" and "right shoe"), i.e., where they attain the status of quasi entry levels within the categorization hierarchy, renders mirror-image discrimination a characteristic feature of visual expertise (Johnson and Mervis 1997; Tanaka and Taylor 1991).

Mirror-image relations between patterns assigned to different categories may affect learning in two different ways: One possibility is that the skill to discriminate between left and right counterparts of mirror-image pairs is acquired via associative learning. This notion can be related to Gross and Bornstein's (1978) hypothesis, according to which mirror images are confused because they are interpreted as two views of one object in three-dimensional space and therefore tend to be linked to the same conceptual node in memory. Alternatively, the tendency to confuse mirror images could arise at the level of stimulus representation as mirror images necessarily share the same local features and therefore produce similar feature descriptions. Learning to distinguish mirror-image pairs would imply a representational shift, during which local features, or isolated pattern parts, are linked to larger entities within a configural description where the symmetry relations between the two patterns can be resolved. Representational shifts from isolated parts to more holistic formats have been proposed as one of the changes that may emerge during the development of expertise in the recognition of faces and other objects (e.g., Farah et al. 1998; Gauthier and Tarr 2002).

The two hypotheses outlined above differ in the way in which they predict generalisation, or transfer, of categorical knowledge involving mirror-image relations. If such relations are mediated by associative learning mechanisms that link stimuli with particular conceptual nodes, then there should be little or no generalization if the same patterns are paired with new labels (nodes) in a subsequent transfer task. In contrast, if learning of mirror-image relations is mediated by representational shifts at the stimulus level, then such shifts, once acquired, should easily transfer to novel tasks in which the same patterns are employed in a different categorization context. We tested these predictions in a category-learning experiment involving a set of 12 Compound Gabor patterns that formed a square-like configuration of four clusters (I–IV) of three patterns each (Fig. 4a,b) within their defining Fourier feature space. Patterns of the cluster pairs I–IV and II–III consisted of mirror images of each other, whereas those of pairs I–II and III–IV did not.

For the first experiment, we combined the four clusters in a pairwise manner in two different ways that either grouped clusters containing mirror images of each other into different categories (condition C1, Fig. 4c left) or into the same category (condition C2, Fig. 4c middle). Two groups of observers were trained

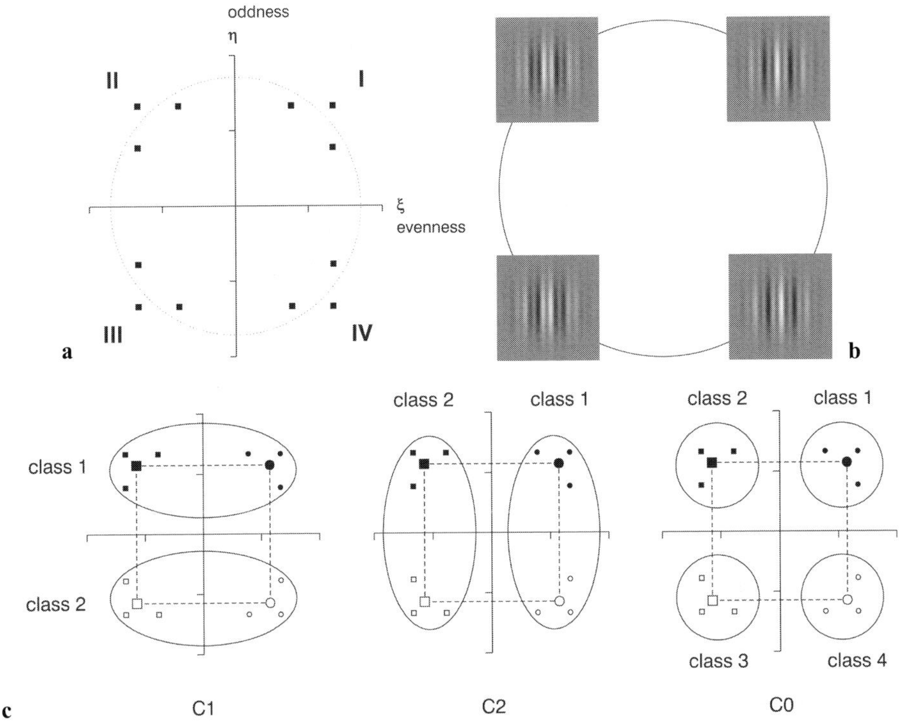

FIG. 4a–c. Experiments on mirror-image relations in category learning. **a** A learning set of 12 iso-energy Compound Gabor patterns was generated, consisting of four clusters I–IV of 3 patterns each *(small symbols)*. Note that the cluster pairs (I, IV) and (II, III) consist of mirror images of each other. Scale: 1 unit = 20 cd/m^2. **b** The four cluster means (not part of the learning set) are illustrated by their grey-level representation. **c** Cluster pairs (condition C1 and C2, Experiment 1) or individual clusters (condition C0, Experiment 2) of the learning set were used to define pattern categories to be learned by the subjects. Note that in the two-class conditions, clusters that were mirror images of each other were either grouped into different classes (C1) or into the same class (C2)

in a supervised learning paradigm in either of these two-class conditions. Once they had reached the learning criterion of a perfect classification they were tested as to the transfer of their conceptual knowledge in a second experiment, where subjects were trained to assign the patterns of each of the four clusters into different categories (condition C0, Fig. 4c right). Thus, the second experiment involved the same stimuli but a different categorization context.

The learning time data of both groups in the two experiments are summarised in Figure 5a. In Experiment 1 (Fig. 5a left), Group 1 (condition C1), which was required to distinguish mirror images during learning, needed an average of 26.2 learning units to reach the learning criterion. In contrast, Group 2 (condition C2), which was not required to discriminate between mirror images during learning succeeded at an average of 2.75 learning units. A complementary data pattern

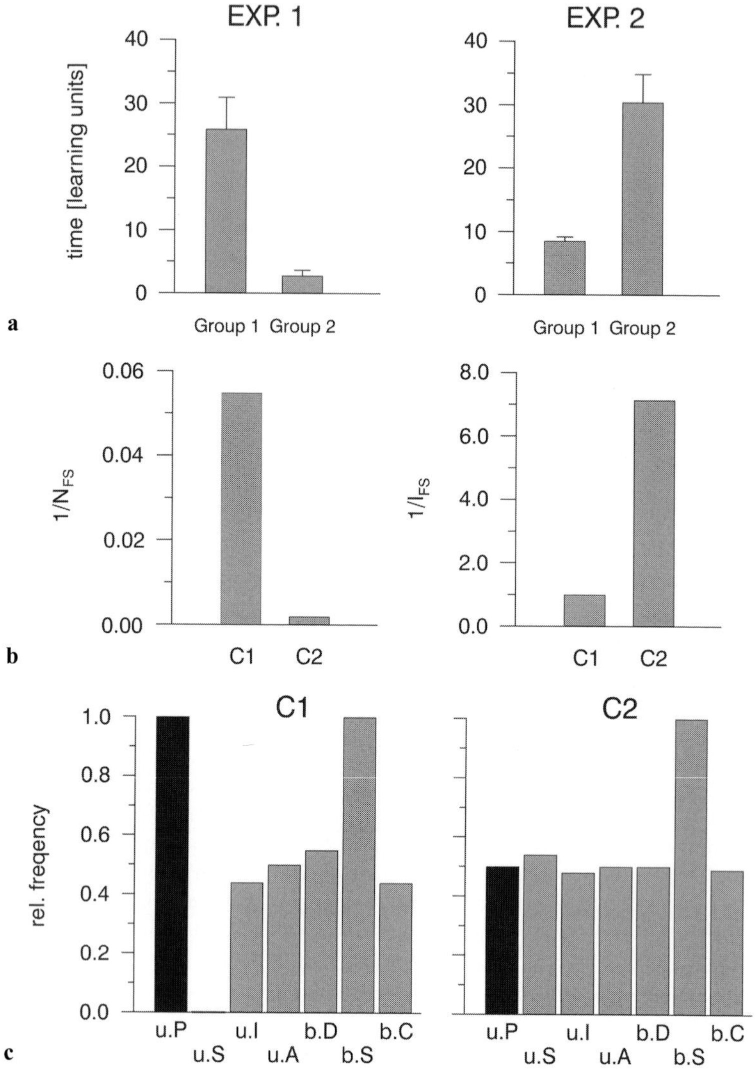

FIG. 5a–c. **a** Mean learning duration (number of learning units to criterion) in Experiment 1 (two-classes configurations C1 and C2) and in Experiment 2 (four-classes configuration C0) for Group 1 and Group 2. **b** EBS-predicted relative learning durations for Experiment 1 (two-class conditions C1 and C2, left) and for Experiment 2 (four-class condition C0, right) with observers being pre-trained in either C1 or C2. Note the complementary pattern of learning times that closely matches the behavioural data shown in (a). **c** Relative frequencies of unary (*u.P*: position, *u.S*: size, *u.I*: luminance, *u.A*: aspect ratio) and binary (*b.D*: distance, *b.S*: relative size, *b.C*: contrast) attributes within the EBS solutions for the classification tasks C1 and C2. Note that the attribute *position* (highlighted in black) attains a predominant role for condition C1, which involves the discrimination of mirror patterns

is found in Experiment 2 (Fig. 5a right): While Group 2 on average learned the patterns after 30.2 learning units, Group 1 required only 8.5 learning units to reach the learning criterion.

The results of Experiment 2 show that subjects who had successfully learned to discriminate between mirror-symmetric counterparts in Experiment 1 generalized this conceptual knowledge to a different categorization context involving the same set of stimuli. This supports the hypothesis that learning to distinguish between mirror images involves a representational shift towards a format in which mirror-image relations are easier to resolve, thus facilitating their integration within categorical knowledge structures. To explore the underlying mental representations, we modelled human performance using an EBS classifier. The simulations followed the same procedure as outlined in section 3 and employed the same system parameters with regard to pattern segmentation, rule generation, and attribute reservoir.

Figure 5b (left) shows the EBS-predicted learning durations for the category learning tasks in Experiment 1. In agreement with the behavioural data, fast learning is obtained for class configuration C2 (cf. Group 2 in Fig. 5a left), whereas slow learning is obtained for class configuration C1 (cf. Group 1 in Fig. 5a left). Figure 5b (right) plots the cross-task compatibility index $1/I_{FS}$, where I_{FS} denotes the number of attribute states that allow the solution of the classification problems with two-classes *as well as* that of the classification problem with four classes (C0, cf. Fig. 4c). This index is *low* for the classification problem C1 and *high* for C2. The latter results explain the complementary pattern observed for the learning duration of Group 1 and Group 2 in Experiment 2 relative to that in Experiment 1 (Fig. 5a).

The relative frequencies of unary and binary attributes within the sets of solutions for conditions C1 and C2 are shown in Figure 5c. These histograms may be regarded as signatures of the underlying categorical representations as they indicate the relative importance of the various attributes within the solutions of the respective classification problems. Accordingly, the signature of C1 differs from that of C2 mainly in that the unary attribute *position* becomes predominant at the expense of the unary attribute *size*. These simulation results further corroborate the representational shift hypothesis by suggesting that the shift crucially implies the use of spatial (positional) information relative to an external (egocentric or allocentric) frame of reference. The use of such positional information permits the unique indexing of parts by their spatial coordinates and therefore indicates a qualitative difference in the internal representation of pattern categories that are acquired in tasks involving mirror-image discrimination.

5 Discussion

Evidence-based classifiers solve a given categorization problem by constructing rules carrying class-specific evidence weights. These rules are based on non-relational and relational attributes of object parts defined in the image domain.

Such strategies can be used to model pattern category learning and generalization to grey-level transformations at the level of the subjects' individual profile of confusion errors (Jüttner et al. 1997). Here we extend these findings by demonstrating that EBS also predicts the relative difficulty of a category-learning task as reflected in learning time. Such simulations allow inferences about the set of attributes evaluated for rule generation, thus providing a "signature" of the underlying conceptual representation. This result has been achieved by first showing how context systematically affects the internal representation of pattern categories (see also Jüttner et al. 2004), and then identifying attributes that are crucial for mirror-image discrimination. Although the experiments involved different sets of patterns, different classification tasks and different types of pattern manipulation to test generalization, the simulations were based on the same set of system parameters. EBS therefore provides, despite its potentially many degrees of freedom, a parsimonious description of human performance.

Evidence-based classifiers provide an explicit link between physical and internal representation, as image segmentation, attribute extraction and rule generation are entirely defined within the (physical) image domain. Such a representational format contrasts with that of standard psychometric approaches to categorization such as prototype models (Reed 1972), exemplar models (Medin and Schaffer 1978; Nosofsky 1986) or General Recognition Theory (Ashby 1989). These models generally represent objects or patterns as single points within a multidimensional psychological space, the metric of which is derived from perceived similarity via multidimensional scaling. A limitation of such models resides in the fact that mapping from physical feature space into psychological space remains algorithmically unspecified. As a consequence, this class of models fails to explain the difficulty of mirror-image discrimination because they remain tacit as to *what* makes mirror stimuli look so similar. The EBS approach solves this problem by relating classification behaviour to a representation based on components that constitute physical pattern structure.

While evidence-based classification adopts a more low-level perspective than traditional psychometric categorization models, it assumes a more high-level stance than physiologically inspired approaches, such as the HMAX model of Riesenhuber and Poggio (1999). HMAX operates directly in image space and consists of alternating layers of linear (S) and non-linear (C) units that perform a hierarchical decomposition of the input image into features defined by the S units. The C units employ a nonlinear maximum operation to pool over afferents tuned to different positions and scales thus achieving invariance to translation and size. Such decomposition could be conceived as a pre-processing front end to an evidence-based classifier to detect the presence of pattern components. However, taken as a stand-alone model the spatial pooling performed by the C units makes HMAX less adequate for categorization tasks involving more complex stimuli, such as mirror patterns, that only differ in the position of their local features. Consistent with this notion, HMAX simulations yield similar confusion patterns for pseudo-mirror views of depth-rotated paper clip objects as observed for neurons in the inferotemporal cortex of the monkey (Riesenhuber and Poggio 1999).

This paper focussed on the application of part-based approaches to pattern categorization but such techniques also provide a promising method of reconciling divergent positions in theories of object recognition between proponents of geon-based models on the one side and view-based accounts on the other. In their original form, geon-based models (Biederman 1987; Hummel and Biederman 1992) provide a framework for object recognition at the level of basic-level categorizations based on structural object descriptions involving non-metric, categorical relations between object components. In contrast, part-based approaches such as EBS capture performance in more complex tasks such as categorization at the subordinate level where continuous, metric relations between object components become crucial for discrimination. However, the effective distinction of attribute values depends on the partitioning used to define rule regions, and only values falling into different partitions activate different rules. When applied to spatial dimensions, such partitioning may result in rules that code discrete-valued, categorical spatial relationships such as "on top of", i.e., categorical relations of the kind that form the reservoir of relational attributes in geon-based accounts. Furthermore, the segmentation algorithm used to extract the part-structure of the input image can be implemented in various ways, for example using Laplacian Filters (Marr 1976) or 2D curvature operators (Zetzsche and Barth 1990). Given this flexibility it is conceivable that the use of components based on non-accidental contour properties, i.e., geons, represents a fast-track route for object recognition at a relatively coarse level of the categorisation hierarchy within a more general part-based recognition system that accommodates sophisticated subordinate classifications.

As an alternative to structural accounts, there is a wide range of so-called image-based models that generally assume that 3D objects are represented in terms of multiple, viewer-centred, two-dimensional (2D) views, among which the visual system interpolates if necessary. Despite this common denominator of viewpoint specificity the understanding of the representational format that constitutes a "view" has changed over the years, from early picture-like representations (Pinker 1984) over those involving simple features like corners or vertices (Poggio and Edelman 1990; Riesenhuber and Poggio 1999) to fragments in more recent versions (Ullman and Sali 2000; Edelman and Intrator 2000). In that sense, view-based accounts have become more structural, which permits linkage to the idea of part-based recognition.

Conversely, part-based recognition schemes allow the implementation of different types of representational formats, depending on whether only relations between adjacent parts are considered, or between all parts. Moreover, representations generated by evidence-based systems in general are "attribute-indexed", i.e., such representations ignore the explicit associations between attributes and pattern parts. While such representations are sufficient for the distinction of objects with differing part structures, these representations necessarily fail in more complex classification tasks, such as the discrimination of mirror images, which are characterized by the same sets of unary and binary attributes. Here attributes need to be associated with the parts to which they refer. In our simulations, this association was re-established by the attribute *position*, which uniquely indexes parts by their spatial coordinates. The resulting representations attain the additional quality of being "part-indexed" and allow for more powerful but computationally more expensive part-based recognition strategies such as graph-matching (see Bunke 2000). From a phenomenological perspective, part-indexed representations can be regarded as one possible realization of a "holistic" format, in which pattern parts become connected to each other in a unique, non-interchangeable way – in contrast to attribute-indexed representations where this uniqueness is not guaranteed. Learning such part-indexed object representations thus implies a shift toward a format in which individual parts form larger constituents akin to the notion of *fragments* proposed in more recent image-based accounts (Edelman and Intrator 2000; Ullman and Sali 2000).

In conclusion, generic part-based approaches provide a unified framework that is general enough to account for recognition at various levels of categorization and sufficiently flexible to accommodate both principles of geon-based and image-based approaches. Indeed, there is recent evidence from behavioural (Hummel 2001; Foster and Gilson 2002; Stankiewicz 2002; Haywood 2003; Thoma et al. 2004) and neuroimaging (Vuilleumier et al. 2002) studies to suggest that view-invariant and view-dependent representations do co-exist in the brain with a relative preponderance that may depend on task, context and level of expertise. Such co-existence would make a single, unified account for the representational format that underlies human object recognition appear a particularly parsimonious theoretical perspective.

References

Ashby FG (1989) Stochastic general recognition theory. In: Vickers D, Smith PL (Eds) Human information processing: measures, mechanisms, and models. Elsevier, Amsterdam, pp 435–457

Biederman I (1972) Perceiving real-world scenes. Science 177:77–80

Biederman I (1981) On the semantics of a glance at a scene. In: Kubovy M, Pomerantz RJ (Eds) Perceptual organization. Erlbaum, Hillsdale NJ, pp 213–253

Biederman I (1987) Recognition-by-components: a theory of human image understanding. Psychol Rev 94:115–147

Binford T (1971) Visual perception by computer. Proceedings, IEEE conference on systems and control. Miami, FL

Bischof WF, Caelli T (1997) SURE: scene understanding by rule evaluation. IEEE Trans Patt Anal Machine Intell 19:1284–1288

Bunke H (2000) Graph matching for visual object recognition. Spat Vis 13:335–340

Caelli T, Dreier A (1994) Variations on the evidence-based object recognition theme. Pattern Recogn 27:1231–1248

Chun MM, Jiang Y (1998) Contextual cueing: implicit learning and memory of visual context guides spatial attention. Cogn Psychol 36:28–71

De Graef P, Christiaens D, d'Ydewalle G (1990) Perceptual effects of scene context on object identification. Psychol Res 52:317–329

Edelman S, Intrator N (2000) Coarse coding of shape fragments) + (retinotopy) ≈ representation of structure. Spat Vis 13:255–264

Fan T, Medioni R, Nevatia R (1989) Recognizing 3-D objects using surface descriptions. IEEE Trans Patt Anal Machine Intell 11:1140–1156

Farah MJ, Wilson KD, Drain M, Tanaka JW (1998) What is "special" about face perception? Psychol Rev 105:482–498

Foster DH, Gilson SJ (2002) Recognizing novel three-dimensional objects by summing signals from parts and views. Proc R Soc Lond B 269:1939–1947

Gauthier I, Tarr MJ (2002) Unraveling mechanisms for expert object recognition: bridging brain and behavior. J Exp Psychol Hum 28:431–446

Gross CG, Bornstein MH (1978) Left and right in science and art. Leonardo 11:29–38

Haywood WG (2003) After the viewpoint debate: where next in object recognition. Trends Cogn Sci 7:425–427

Henderson JM, Weeks PA, Hollingworth A (1999) The effects of semantic consistency on eye movements during complex viewing. J Exp Psychol Hum 25:210–228

Hummel JE (2001) Complementary solutions to the binding problem in vision: implications for shape perception and object recognition. Vis Cogn 8:489–517

Hummel JE, Biederman I (1992) Dynamic binding in a neural network for shape recognition. Psychol Rev 99:480–517

Jain AK, Hoffman R (1988) Evidence-based recognition of 3-D objects. IEEE Trans Patt Anal Machine Intell 10:783–802

Johnson KE, Mervis CB (1997) Effects of varying levels of expertise on the basic level of categorization. J Exp Psychol Gen 126:248–277

Jüttner M, Rentschler I (1996) Reduced perceptual dimensionality in extrafoveal vision. Vision Res 36:1007–1022

Jüttner M, Caelli T, Rentschler I (1997) Evidence-based pattern recognition: a structural approach to human perceptual learning and generalization. J Math Psychol 41:244–258

Jüttner M, Langguth B, Rentschler I (2004) The impact of context on pattern category learning and representation. Vis Cogn 11:921–945

Marr D (1976) Early processing of visual information. Philos T Roy Soc B 275:483–524

Marr D, Nishihara HK (1978) Representation and recognition of the spatial organization of three-dimensional shapes. Proc R Soc Lond B 200:269–294

Medin DL, Schaffer MM (1978) Context theory of classification learning. Psychol Rev 85:207–238

Morris RK (1994) Lexical and message level sentence context effects of fixation times in reading. J Exp Psychol Learn 20:92–103

Nosofsky RM (1986) Attention, similarity, and the identification-categorization relationship. J Exp Psychol Gen 115:39–57

Palmer SE (1975) The effects of contextual scenes on the identification of objects. Mem Cogn 3:519–526

Pearce AR, Caelli T (1999) Interactively matching hand-drawings using induction. Comput Vis Image Underst 73:391–403

Pinker S (1984) Visual cognition: an introduction. Cognition 18:1–63

Poggio T, Edelman S (1990) A network that learns to recognize three-dimensional objects. Nature 343:263–266

Reed SK (1972) Pattern recognition and categorization. Cogn Psychol 3:382–407

Riesenhuber M, Poggio T (1999) Hierarchical models of object recognition in the cortex. Nat Neurosci 2:1019–1025

Rivlin E, Dickenson S, Rosenfeld A (1995) Recognition by functional parts. Comput Vis Image Underst 62:164–176

Rosch E (1978) Principles of categorization. In: Rosch E, Lloyd B (Eds) Cognition and categorization. Erlbaum, Hillsdale NJ, pp 27–48

Shapiro L, Haralick R (1981) Structural descriptions and inexact matching. IEEE Trans Patt Anal Machine Intell 3:504–519

Stankiewicz BJ (2002) Empirical evidence for independent dimensions in the visual representation of three-dimensional shape. J Exp Psychol Hum 28:913–932

Tanaka JW, Taylor M (1991) Object categories and expertise: is the basic level in the eye of the beholder? Cogn Psychol 23:457–482

Thoma V, Hummel JE, Davidoff J (2004) Evidence for holistic representations of ignored images and analytic representations of attended images. J Exp Psychol Hum 30:257–267

Ullman S, Sali E (2000) Object classification using a fragment-based representation. In: Lee SW, Bülthoff HH (Eds) Biologically motivated computer vision. Springer, Berlin Heidelberg New York, pp 73–87

Vuilleumier P, Henson RN, Driver J, Dolan RJ (2002) Multiple levels of visual object constancy revealed by event-related fMRI of repition priming. Nat Neurosci 5:491–499

Zetzsche C, Barth E (1990) Fundamental limits of linear filters in the visual processing of two-dimensional signals. Vision Res 30:1111–1117

5
Recent Psychophysical and Neural Research in Shape Recognition

Irving Biederman

1 Introduction

Shape is the major route by which we gain knowledge about our visual world. All contemporary theories of shape-based object representation, e.g., Hummel and Biederman (1992); Riesenhuber and Poggio (2002), assume a hierarchy of features by which the initial Gabor-like filtering that is characteristic of V1 cell-tuning is ultimately transformed through a series of stages to a point where cell tuning is better described by "moderately complex" features with receptive fields (r.f.s) that often cannot be analyzed into their linear components (Tanaka 1993; Kobatake and Tanaka 1994). These later stages are the inferior temporal cortex in the macaque (IT) and, in humans as determined by fMRI, likely the Lateral Occipital Complex (LOC). Along with the increase in r.f. nonlinearity in IT and LOC, the cells exhibit a high degree of invariance to changes in the conditions of presentation so the response is only moderately changed to variations in the viewing conditions. In this chapter we will review recent evidence, both behavioral and neural, that shed light on the nature of these object representations.

A bit of recent history and commentary on that history.

Theories of object recognition are often termed "controversial," particularly in accounting for the effects of rotation in depth. The apparent controversy has been phrased in terms of whether vision is "view-based" (e.g., Poggio and Edelman 1990) or "invariant." But what is the controversy? As we have noted previously (Biederman and Bar 2000), *all* accounts of vision have to be view-based. The alternative is Extra Sensory Perception (ESP)!

The empirical issue has been defined in terms of whether there is a zero *vs* some cost in matching or recognition of an object when it is presented at an orientation in depth other that the orientation of its initial presentation. Again, all accounts of vision would have to say that under most circumstances there should be *some* cost. In an extreme case, one cannot know what the back of a house looks like from looking at its front, aside from generalization from the viewing of similar

Departments of Psychology and Neuroscience Program, Hedco Neurosciences Building, MC2520, University of Southern California, Los Angeles, CA 90089-2520, USA

houses. But what about the quite common intermediate case where *some* of the perceptual information from one view can be discerned from another? In my 1987 *Psychological Review* paper, I proposed a "geon recovery principle" by which a similarity function could be computed, akin to Tversky's (1977) aspects of similarity measure, that was a positive function of the number of geons that were present in both views and a negative function of the geons that were present in view A but not in view B and the number of geons that were present in view B but not in view A. There would have to be weightings of these geons in terms of their perceptibility (resolution) due to foreshortening and self-occlusion as well as their diagnosticity to the object/response class, etc. I also speculated that the same function might define the similarity between any two objects or an intact object and a version missing some parts, etc. There is considerable evidence that the overlap in parts does predict, at least qualitatively, rotation costs (e.g., Biederman and Gerhardstein 1993) but a full quantitative account is still lacking. It is unfortunate, in my opinion, that so much ink has been spilled on attacking a position – zero rotation costs for all conditions of rotation – that no one ever held.

Insofar as I noted that different representations (i.e., different GSDs) could be required for substantially different views (Biederman 1987), what strikes me as particularly ironic is that I probably should be regarded as the (an?) originator of modern versions of view-based theories!

Wallraven and Bülthoff's (this volume) "view-based" account assumes that the alternative to their position is a 3D model, of the kind proposed by Marr (1982). But there is no doubt that Marr, whose account of object recognition was admittedly tentative, would not (indeed, cound not,) argue that one could know what the back of a house looked like from just seeing its front (igloos excepted). They ascribe to me a similar account, but as noted previously, Biederman (1987) was quite clear that different GSDs would be required for large differences in viewpoints. What they seem to have missed is what many have been arguing now for 20 years: Nonaccidental properties, available from a 2D image, offer an extraordinarily powerful way to achieve 3D view invariance – as long as the surfaces that project these properties are available in the image.

2 Four Issues: Invariance, Structural Descriptions, Nonaccidental Properties, and Surface Features

In this section I will consider some of the outstanding issues and then later briefly review research, both psychophysical and neural, relevant to these issues.

2.1 Invariance

Given that all contemporary theories posit a hierarchy of features – rather than, say, the direct matching of Gabor kernels – what are the issues that remain? One, of course, is how invariance can be achieved. An object seen at one position, orientation in depth, direction of illumination, and size, can be often recognized with little or no cost when seen at another viewpoint where these variables can undergo considerable changes in their values. The Riesenhuber and Poggio (2002) model achieves a degree of invariance, as does the Hummel and Biederman (1992) network. I will only consider invariance to rotation in depth in this chapter.

2.2 *Nature of the Representation: Feature Hierarchies* vs *Structural Descriptions*

As noted previously all theories posit a hierarchy of features by which Gabor-like units with local r.f.s are ultimately mapped unto highly non-linear units with a high degree of invariance. An excellent example of non-linear units can be found in the V4 L-vertex units discovered by Pasupathy and Connor (1999). An L vertex, as its name implies, is formed by the cotermination of two (non-collinear contours) at a common point. The V4 L-vertex units are each tuned to a particular angle (e.g., 60 deg.) and a particular orientation (e.g., with the bisector vertical) of the vertex. These units are nonlinear in that they are neither activated by the bisector of the angle nor by a single leg of the vertex. Both legs are required. But is a set of features consisting of vertices and lines sufficient for understanding object recognition?

Some accounts (e.g., Riesenhuber and Poggio 2002) would answer this question in the affirmative. Indeed, their Hmax model does an impressive job in assigning new object instances into previously learned object categories.

An alternative theoretical approach also assumes a feature hierarchy but maps the features onto a *structural description* (SD) (Humphreys and Riddoch 1987; Biederman 1987; Winston 1975). A structural description distinguishes parts and relations. It thereby allows reasoning about objects so that not only can the model determine that two images represent different objects, but *how* they differ. For example, given two objects, one with a cylinder on a brick and the other with a wedge on an identical brick, we can readily perceive that it is the top parts of the two objects that differ in shape, even if the bricks were not aligned horizontally.

As noted above, a major distinction between feature hierarchies and S.D.s is that relations are explicitly defined and distinguished from parts in an S.D. but not in a feature hierarchy. Instead the relations in a model such as Hmax are implicit in a 2D coordinate space. A set of features (which in a S.D. might correspond to a part), might have coordinates that, if read out explicitly (as they are in a S.D.), could show that these features were "above," or "larger than," or "connected end-to-end" with another set of features but those labels, e.g., "above," "connected end-to-end," don't exist in Hmax.

2.3 *Nonaccidental* vs *Metric Properties*

Geon theory is a particular instantiation of structural descriptions (i.e., geon structural descriptions, GSDs) (Biederman 1987; Hummel and Biederman 1992). GSDs place heavy reliance on nonaccidental properties (NAPs). NAPs are qualitative properties of (in the case of shape) orientation and depth discontinuities, which are largely unaffected by rotation in depth. For example, whether a contour is straight or curved is unlikely to change as the object rotates in depth. In contrast to NAPs, much image variation is metric (MPs, for metric properties), such as degree of curvature or the length of a contour. Whereas small differences in MPs are registered with difficulty, differences in NAPs provide a ready basis for distinguishing one object's parts and relations from another (e.g., Biederman and Bar 1999). Neither a classification of contour by NAPs nor explicit parts nor explicit relations are specified by view-based templates or current feature hierarchy accounts.

2.4 Surface Features vs *Orientation and Depth Discontinuities*

Still another distinction between Geon Theory and Hmax is, as noted above, that Geon Theory extracts the shape of an object as defined by its orientation and depth discontinuities. Hmax just takes the image as is – surface characteristics such as color and texture as well as the object's shape. Thus geon theory would tend to minimize the differences between a photograph of an object and its line drawing rendition. For Hmax, this would be an enormous difference.

3 Object Reasoning

To illustrate what is meant by object reasoning, imagine performing a matching task in which you are to determine if two sequentially presented novel objects are the same or different, irrespective of their orientation in depth. Before scrutinizing Figure 1, please cover the objects with your hand. The figure illustrates some possible trials in which the object on the left is always S1, the first object. The objects in the right column are possible S2s. Take a quick peek at S1. You probably can describe it. Now take a quick peek at the top object in the S2 column. It should be trivially easy to respond "different." The same would be true of the second object in the S2 column. Or the fourth. You might judge the third object to be "same," even though the object is now rotated in depth, as is the wedge in the previous object. The first three trials differed in at least one geon and the discrimination is trivially easy. The last object had the same geons and it looks the same as S1, despite its depth rotation. Little or no rotation costs would be expected with such objects.

4 Empirical Research

4.1 NAPs (Geons) vs *MPs*

GSDs specify both parts and relations. I will here concentrate on the NAP characterization of the parts (see Biederman 1995, for a summary of the evidence supporting the role of simple parts in the representation mediating visual object recognition).

The benefit conferred by NAPs, documented by Biederman and Gerhardstein (1993), and confirmed by Biederman and Bar (1999) is quite dramatic and is among the largest effects in shape recognition. This benefit does not appear to depend on exposure to regular, simple artifacts that are so prevalent in environments in the developed world. Recently Nederhouser et al. (2005) reported that the Himba, a semi-nomadic tribe in northeastern Namibia with minimal exposure to developed-world artifacts, showed markedly better performance in a match-to-sample task in matching single geons when the distractor differed in a NAP than an MP. In fact, the NAP advantage for the Himba was identical to that shown by University of Southern California undergraduates, suggesting that the connectivity subserving the NAP advantage develops from robust statistics that would hold with virtually any natural environment.

IT tuning also show greater sensitivity to differences in NAPs compared to MPs. Kayaert et al. (2003) showed the IT cells in the macaque modulated their firing more to a change in a NAP compared to a change in an MP when the differences in NAPs and MPs were equated by a measure of pixel energy.

4.2 Matching Depth-Rotated Objects

Distinctive NAPs can confer an enormous benefit in attempting to determine whether two bent paper clips are the same or different when they are viewed at different orientations in depth (Biederman and Gerhardstein 1993). These investigators substituted a different geon for each center segment of a set of 10 line drawings of bent paper clips. The addition of the distinctive geon dramatically reduced rotation costs (to 5,000/sec) from a level with error rates so high that RTs were, essentially, uninterpretable.

View-based template accounts, in assigning no special status to NAPs or parts, would require familiarity with the specific views of novel objects, with only a modest generalization gradient around a nearby view. Some (Tarr and Bülthoff 1995) protested this demonstration, arguing that NAPs were of value only with a small set of known stimuli where people could anticipate a distinguishing NAP. That people would spontaneously exploit distinguishing NAPs was, indeed, one of the points that Biederman and Gerhardstein wished to make, but is familiarity required to get immediate viewpoint-invariance with novel objects?

Moshe Bar and I (Biederman and Bar 1999) compared directly the rotation costs for detecting differences in either a Metric Property (MP) or a NAP in a same-different sequential matching task. We used novel, rendered two-part objects, such as those shown in Figure 1, presented at either the same or different orientations-in-depth. On half the trials the objects were identical; on half the trials they differed in either an MP, e.g., aspect ratio, of a single part or a NAP, e.g., straight *vs* curved axis (producing a different geon) of a single part. The contrast of the object on the left and the third object in the right hand column of Figure 1 illustrates a NAP difference (straight- *vs* curved-axis cylinder). The MP variation would have been a cylinder with a different length (aspect ratio) or angle of attachment to the wedge. The subjects saw a given

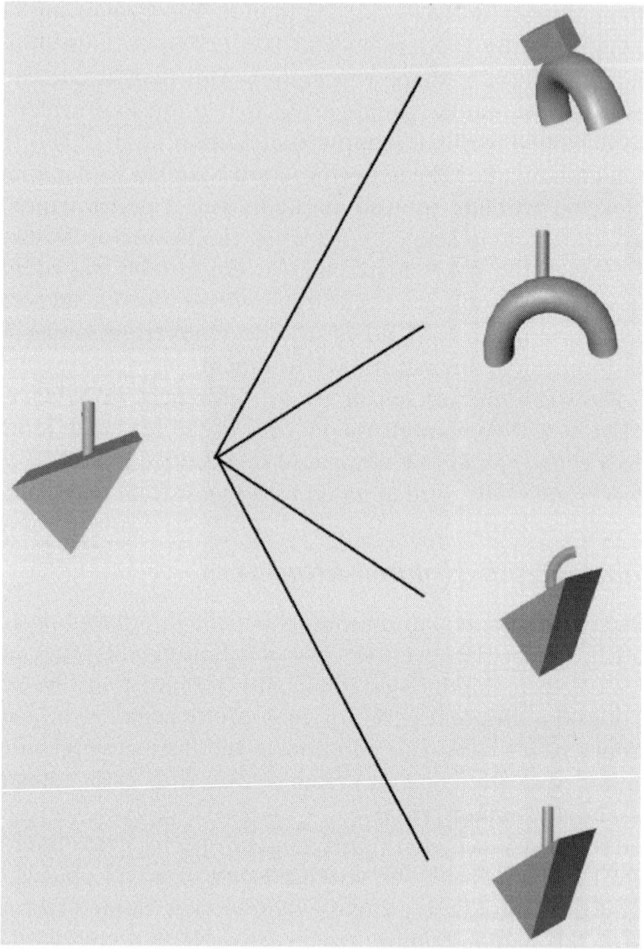

FIG. 1. An illustration of four trials in a Same-Different matching task of two-geon novel objects (from Biederman 2000). The object on the left is S1, the first stimulus for all four trials. The four objects in the right column are possible S2s. The top object differs in both geons; the second and third in one geon, and the bottom object is the same, but rotated in depth. Observers should have no trouble accurately performing same-different judgments. Nor should they have any difficulty in describing the objects and how they differ from each other. Only the third and fourth S2s would have been trials in the Biederman and Bar (1999) experiment

stimulus sequence only once, so they could not predict whether a part would change, or, if there was a change, which part would change and how it would be changed.

How much of an MP difference is equivalent to a given NAP difference? This apples-and-oranges problem is critical for any principled answer to this question. In the Biederman and Bar (1999) experiment the MP and NAP differences were selected to be equally discriminable, as assessed by RTs and error rates, when the objects were at the same orientation in depth (0° orientation difference). The MP image differences were also approximately the same magnitude (actually slightly larger) then the NAP image differences when the images were scaled according to a wavelet-like similarity measure (Lades et al. 1993). Rotation angles that averaged 57° produced only a 2% increase in error rates in detecting the NAP differences but a 44% increase in error rates (to a level that was below chance) in detecting MP differences.

Rotation costs, though small, are often apparent even with distinguishing GSDs (Biederman and Gerhardstein 1993; Biederman and Bar 1999; Hayward and Tarr 1997; Tarr et al. 1997; Tarr et al. 1998). What might be producing these costs? It is possible, as noted by Biederman and Gerhardstein (1993), that an orientation-specific representation underlies these costs. This representation may be of one of two types, given current theorizing: a) an episodic representation that binds view information along to an invariant representation of shape, as detailed by Biederman and Cooper (1992), see below), that could be employed on some percent of the trials to mediate performance (though not necessarily object perception), or b) that there are viewpoint specific representations directly mediating object perception. But before the latter alternative is accepted merely on the basis of some costs with distinguishing GSDs present, other possible bases for the costs must be ruled out.

Look again at Figure 1 and consider what one would have to do to make it difficult, under rotation, to determine that the third S2 object was different from the first and the fourth S2 object was the same. One way would be to render the object in such a way that it would be difficult to determine if the distinguishing geon was curved or straight. Biederman and Bar (1999), in a critical review of those studies reporting high rotation costs, noted that low resolution of the distinctive geon was a common characteristic in such studies. Biederman and Bar (1998) showed that factors that increased the discriminability of distinguishing geons in rendered images, such as avoiding near accidents or using increased exposure durations, greatly reduced rotation costs.

There is another, subtler, effect that could have contributed to the apparent costs of rotation in the Tarr et al. (1997, 1998); Hayward and Tarr (1997) same-different matching studies. On rotated Same trials and all Different trials in a Same-Different matching task, a "difference" signal might be produced by the change in luminance of specific display positions. This signal may be related to Nowak and Bullier's (1997) finding that marked changes – a transient – in a stimulus produce a signal that rapidly propagates through the ventral pathway all the way to frontal cortex. (Because of the intervening mask, the difference signal would be between S2 and an actively maintained representation of S1, as noted by Biederman and Bar 1999.) No difference signal would be produced when S1 and S2 are the same, unrotated, object in the same position. So the subject could readily use the *absence* of a difference signal to respond Same on unrotated (0°) trials, artifactually lowering reaction times (RTs) on such trials with the consequence that the slope of the RT X Rotation Angle function would be increased. Biederman and Bar (1998) showed that the effect of this artifact in increasing rotation costs could be greatly reduced by merely shifting S2 with respect to S1 on all trials, so that the difference signal was always present. The translation, by producing a difference signal on all trials, served to raise the RTs and error rates for 0° trials relative to rotated trials. This had the effect of greatly reducing the *apparent* costs of rotation. Biederman and Bar's (1999) experiment, which found near invariance over rotation, also translated S2 with respect to S1 on all trials.

4.3 Observations About Bent Paper Clips as Experimental Stimuli

Many of the studies documenting large rotation costs have employed stimuli resembling bent paper clips. The central motivation for devising such stimuli (and others of similar design) was that they would be unfamiliar, so that the learning of different poses could be studied. However, one must consider an obvious characteristic that accounts for much of the extraordinary difficulty in classifying members of sets of such stimuli: The members of such sets are not distinguished by GSDs.

The absence of distinguishing GSDs in the standard set of bent paper clips means that the critical information for everyday shape recognition is absent from these stimuli so the relevance of such objects to normal recognition can be questioned. Some bent paper clip devotees have suggested that their stimuli are relevant for subordinate-level recognition, such as the difference between different kinds of tables. However, a review of the vast majority of subordinate-level classifications that people make in their lives suggests that it is extremely rare that distinguishing GSDs are not available. A square table can be distinguished from a round table without appeal to metric information and certainly without engaging in mental rotation. Biederman et al. (1999) note that NAPs of small regions, rather than metric templates, are specified for discriminating among highly similar classes such as birds on the same page in the bird guides.

Think of how you would discriminate two different chairs of the same manufacturer's model. Without fail, visitors to my office look for a distinctive scratch or stain or other such *nonaccidental* difference, at a small scale. They never consider what is readily expressed by metric templates – a template of the whole chair or, in selecting a small feature, those that might differ metrically (at a modest scale).

The objects shown in Figure 1 meet the criteria of being unfamiliar, yet in retaining distinctive geons they allow one to study how such information might be used. Although a set of paper clips lack distinctive GSDs, their projections often provide an accidental or near accidental characteristic that people try to interpret in terms of GSDs (Biederman and Gerhardstein 1993, 1995; Biederman and Bar 1998). For example, the bottom S2 object in Figure 2 resembles an arrow that would normally be produced by actual cotermination of segments but is here an accident of viewpoint. Biederman and Bar (1998) observed that when there were such qualitative differences in appearance – typically well captured by differences in GSDs – miss rates were extremely high. When S1 and S2 were actually

FIG. 2. Illustration of four trials in a Same-Different matching task (from Biederman 2000) for bent paper clips of the kind that could have been run by Edelman and Bülthoff (1992). Only the bottom S2 is identical to S1 (but rotated in depth)

different clips but with similar GSDs, as in the upper three S2s of Figure 2, then the false alarm rates were extremely high.

As rotation angles increase from 0 to 90°, there is an increasing chance of changes in the qualitative characterization of such stimuli. The oft reported increase in matching costs with increasing rotation angles may be more a consequence of an increasing chance for a change in an accidental GSD then in the rotation of a template. Consistent with this interpretation are the low rotation costs for 180° rotations. Such rotations often approximate mirror reflections under which the GSDs are preserved.

4.4 When GSDs are Insufficient

There is no doubt that aspects of early cortical representation are well described by a 2D array of local filters at a variety of scales and orientations. The view expressed here is that the outputs of such a representation are mapped onto nonaccidental classifiers – such as units distinguishing straight from curved lines or various vertices produced by cotermination of end stopped activity. A vector representing the activity of these nonaccidental classifiers (which, in JIM, are bound through correlated firing), in turn, activate units akin to Hummel and Biederman's (1992) geon feature assemblies, representing single or pairwise combinations of geons and their invariant nonaccidental relations, such as: VERTICAL_CYLINDER_ABOVE_PERPENDICULAR_SMALLER_THAN_X. The output of such geon feature assemblies could readily map onto language, as evidenced by the manner in which people describe the objects in Figure 1, as well as memory structures supporting object cognition.

What if the stimulus does not have distinguishing parts and nonaccidental properties, as with the set of smooth blobby shapes studied by Shepard and Cermak (1973)? In such a case the nonaccidental classifiers would not be differentially activated to distinguish the members of the stimulus set and the observer would have to rely on whatever metric information distinguished the stimuli, in which case the similarity space would be that established by the early local, multiscale, multioriented Gabor-like filters (Biederman and Subramaniam 1997). It should also be the case that discrimination among such stimuli should be more difficult than if distinctive GSDs were available (at the same level of spatial filter similarity), show more rotation costs, be difficult to articulate, and not be the basis of natural concept distinctions.

Discrimination performance among a set of highly similar faces shows such characteristics (Biederman and Kalocsai 1997), as well as similar pairs of the Shepard and Cermak (1973) shapes (Biederman and Subramaniam 1997) and objects with irregular parts (Cooper et al. 1995). See Biederman (1995) for a review.

5 Can View-Based Accounts Incorporate Geons as a Unique or Diagnostic Feature?

Given my earlier point that "view-based accounts assign no special status to NAPs," one can ask whether view-based theorists could regard geons as some kind of unique or diagnostic feature extracted from a 2D view. The answer is, "of course." But there is a serious problem with an account that holds that a unique or diagnostic feature will be employed for recognition. How does the perceiver know what is unique or diagnostic the first time he or she views an object? Consider, again, an individual seeing the nonsense object on the left side of Figure 1. The coding of that object would, roughly, appear to be a vertical cylinder on top of a wedge. That is, the object is described in terms of its simple parts and the relations among these parts (Tversky and Hemenway 1984). This type of representation, a geon structural description (GSD), may well be the default description that the visual system generates in the absence of explicit knowledge about the other to-be-discriminated objects. GSDs often convey the functionality – or affordances – of the object. Moreover, GSDs often readily map on to verbal descriptions and allow reasoning about objects. We can readily say how the four objects on the right side differ from the one on the left (or from each other).

The important question is not whether a representation is view-based, but what that representation is (as, again, *all* representations are view-based). The phenomena of: a) small rotation costs with distinctive GSDs when matching novel objects, b) the sizable costs in recognizing new views of objects, such as a set of bent paper clips, that are not distinguished by GSDs (as discussed in the next section), and c) the reduction in the costs in b) from learning the new views, has obscured the issue of representation, insofar as the nature of what was learned was often not considered. In allowing translational and scale feature invariance, the recent Riesenhuber and Poggio (2002) scheme resembles an earlier proposal by Bartlett Mel (1997). There is nothing in the Riesenhuber and Poggio model to suggest the enormous inferential leverage and invariance to rotation costs provided by distinctive NAPs or the difference in recognizability between recoverable and nonrecoverable contour deletion. These models are, essentially, feature lists in that they do not posit explicit structures, such as parts and relations among parts, by which objects might be represented – and described. Instead, different arrangements of the parts merely produce new features. A potential serious shortcoming of such models is that it is not clear how well they would do with novel objects that are to be distinguished from unknown sets of other objects, such as with the task illustrated in Figure 1.

6 Recent Neural Evidence for GSDs

6.1 Parts in IT

It has long been known that macaque inferior temporal (IT) neurons are highly shape selective and that different neurons show different shape preferences. Tanaka (1993) demonstrated that these preferences could be elicited quite strongly to features of "moderate complexity," typically composed of one or two parts. This level of complexity closely matches what would be expected from single geons, invariant shape features, and, most frequently, geon feature assemblies (Hummel and Biederman 1992), in which two geons are bound in a specific relation.

What occurs in IT when a macaque views an object? Tanifuji and his associates (Tsunoda et al. 2001) have employed optical imaging to address this question. Viewing a complex object such as a fire extinguisher does not activate the whole region homogenously. Instead, several "spots" of activity are apparent. This group then recorded the activity of individual neurons within these spots, using Tanaka's (1993) reduction technique in which parts of a complex stimulus are removed in an effort to determine the specific feature(s) driving the cell. For the most part, neurons within a spot tended to respond to a single part of the object, such as the hose or the barrel, without any reduction in activity compared to their response to the complete object although, occasionally, the removal of parts of the object resulted in increased firing, suggesting inhibition of that neuron from the removed part. A neuron responding to the curved hose cease to fire when the hose was straightened consistent with a general finding that to a first approximation, Tanaka's (1993) and Kobatake and Tanaka's (1994), moderately complex features are viewpoint invariant. Consistent with this interpretation is Esteky and Tanaka's (1998) results showing that metric variation, viz., changes in aspect ratio that would be produced by a rotation in depth, had only a minimal effect on IT cell activity.

6.2 NAPs vs *MPs in IT*

Vogels et al. (2001) tested macaque IT (area TE) neurons with the identical set of two-geon stimuli used by Biederman and Bar (1999) to determine if greater modulation in cell activity would be produced by a change in a geon compared to a change in an MP (Metric Property). They found that geon changes, despite their smaller image changes (as assessed by wavelet similarity measures), produced greater modulation (up or down) in cell activity. Moreover, when the original objects were rotated (i.e., those without an MP or geon change), the modulation attributable to the rotation itself was highly correlated with the modulation produced by MP changes for that cell but completely independent of the modulation produced by the geon changes. Such a tuning pattern would be expected given the results of Biederman and Bar (1999) that only geon-changed stimuli were readily discriminable from the originals under rotation.

As noted earlier, Kayaert et al. (2003) more recently replicated the Vogels et al. (2001) effect of greater modulation from NAP as compared to MP changes. They scaled the image changes by a pixel energy measure. MP image changes had to be approximately 50% larger than NAP changes to produce the same degree of modulation. Moreover, the amount of modulation produced by depth rotation was equivalent to the modulation produced by nonrotated MP changes when the two conditions were equated according to the magnitude of image change.

6.3 Recent Neural Results Supporting a Geon Account of Shape Representation

Three assumptions of geon theory are that the representation of parts a) tends to be simple, b) that the geons are a partition of the set of generalized cylinders based upon nonaccidental differences in the generating function by which a cross section is swept along an axis, and c) that the information can be derived from orientation and depth discontinuities of the original image. Results from recent single-unit studies provide strong support for these assumptions. In the Kayaert et al. (2005b) investigation, macaques passively viewed 2D regular (i.e., simple) and irregular shapes while neurons in area TE were recorded. A difference in a regular shape, say between a circle and a square, produced markedly more absolute modulation (i.e., change in firing, up or down) than a change in a highly irregular shape, where the two types of changes were matched with respect to pixel similarity. If the irregular shapes differed in a NAP (viz., with round *vs* straight contours), then the cells modulated more, suggesting that the sensitivity to NAPs can be witnessed even with irregular shapes.

Kayaert et al. (2005a) discovered that a population code of IT neurons represents independent dimensions of generalized cylinders. For example, a given cell might respond to a highly curved axis of shape independently of its taper, aspect ratio, or curvature of its sides. These cells were, to a great extent, tuned to one end of a dimension or the other, with very few cells preferring intermediate values. Thus a cell could respond predominantly to a highly curved axis while another cell would respond to a straight axis with the firing declining as the axis curvature was changed away from the maximally preferred value.

In the Kayaert et al. (2003) study demonstrating greater sensitivity of IT cells to NAP compared to MP changes, the preferences were unaffected by depicting the shapes as 3D volumes, 2D silhouettes, or line drawings. This suggests that the shape preferences are tuned to the orientation and depth discontinuities. Consistent with this result is a finding by Kourtzi and Kanwisher (2000) that adaptation of the fMRI BOLD signal when viewing a sequence of two object images is maintained when the image changes from a grey-level photograph to a line drawing of the same object. A change in the object causes a release of the adaptation, i.e., a larger bold signal. The lack of an effect of image variables – and a form of invariance – was demonstrated by Vogels and Biederman (2002) who showed that the preferences of IT cells to rendered 3D objects was largely maintained irrespective of changes in the direction of illumination, changes which produced large effects on the image itself.

6.4 Familiarity

There have been a number of reports of TE cell preferences reflecting experimental manipulations of familiarity (e.g., Logothetis et al. 1994; Tanaka 1996). There are two points to be made about such demonstrations. First, tens, if not hundreds, of thousands of trials are required to obtain such preferences (Logothetis 1999). Second, as discussed previously, it is not unlikely that there are at least two functions of object recognition subserved by IT. One is to provide descriptions of objects, novel or familiar, such as what the reader experienced when first viewing S1 in Figure 1. Such a system is likely well developed by late infancy. The second function is to provide an episodic record of the perceptual experience with particular objects or scenes. It is possible that the cells found in the training experiments are those subserving this second episodic memory function. That there may be these two representations of objects was documented by Biederman and Cooper (1992) who showed that the priming of object naming was invariant with size changes but that such changes produced considerable interference on episodic old-new judgments of the shape of the object (in which size was to be ignored). Distractors in that experiment were objects with the same name but a different shape. Similar results were found for changes in position and reflection (Biederman and Cooper 1991) and orientation (Cooper et al. 1992). Although the first function probably supports lion's share of human object

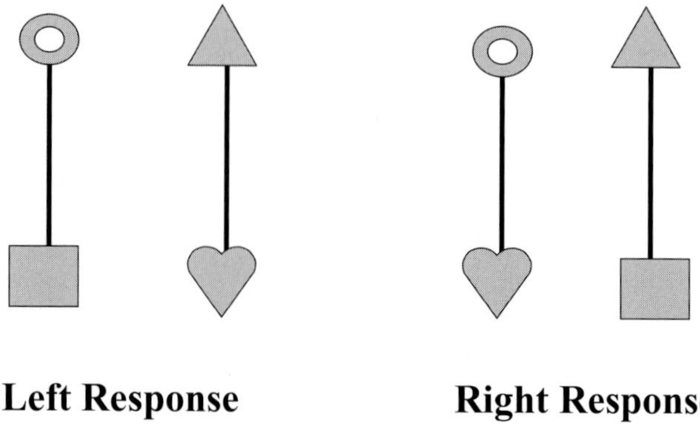

FIG. 3. Illustration of the stimuli from the Baker et al. (2002) experiment. On each trial the monkeys saw one of the four batons. The two left batons were assigned to the left key; the two right to the right key. Note that the individual shapes and their positions (top or bottom), by themselves, are insufficient to determine the correct response. The monkey must process the conjunction of the two shapes. After about 28,000 training trials, cells in IT were found that responded to the individual batons but three times as many were found that responded to the individual shapes

recognition, it would certainly be possible to employ the second problem to solve particular classification tasks. If I know that a chair is on the right and a table on the left, a flash of an object on the right could lead me to infer that it was a chair rather than a table. Such view information could be employed whenever there was difficulty in determining an object's GSD.

Baker et al. (2002) trained macaques to classify four vertical "batons," each with one shape on top and another shape on the bottom, into two classes. The assignment of batons to responses was such that the monkey could not use an individual shape to make a response (see Fig. 3). For example, one of the two batons assigned to the left key could have a circle on top and a square on bottom while the other would have a triangle on top and a star on the bottom. The two right batons would be one with a circle on top and a star on the bottom while the other had a triangle on top and a square on bottom. After about 25,000 trials cells in IT were found that responded to the individual baton but three times that many still responded to the individual shapes, irrespective of what else was assigned to it. These results indicate that specific combinations of features can be learned but that the dominant coding in IT seems to be the individual shape.

6.5 Structural Descriptions

Despite the common assumption of structural descriptions in cognition and their value in object reasoning as described previously, there has been, until recently, no direct evidence for them. Behrmann et al. (2006) have recently reported a patient with a lesion of the left LOC (approximately) who, at first glance, is sensitive to the shape of parts but not at all to the relations among these parts. After learning four two-geon objects he was able to determine when a geon changed but was completely insensitive to a change in the relations. However the patient presents simultagnosia so it is possible that he can only process a single geon at a time.

Hayworth and Biederman (2005) reported an fMRI study in which subjects viewed brief two-frame "flip movies" in which one part of a two-geon object cycled between two different shapes so that a cylinder on top of a brick could change into a pyramid and back again for several cycles. A 24-sec block of trials consisted of three of such geon change movies with the particular shape change varying between movies. In another block the geon would retain its shape but vary its relations, such as the cylinder moving from vertically on top of a brick to horizontally to the side of the brick. The magnitude of these image changes were equated with respect to pixel energies and, indeed, MT was equally affected by the different kinds of changes. For every subject for every voxel in LOC, greater activity was associated with a change in part shape compared to a change in the relations between parts. In fact the relations condition did not lead to greater activity than a control condition in which the object retained its shape but merely rotated in depth. However a region of the intraparietal sulcus showed markedly greater activity to the relations condition than the part shape condition.

7 Conclusion

The evidence suggests that GSDs provide a suitable representation with which to understand the large body of results that have recently accumulated in the psychophysical study of depth-rotated objects as well as single unit and fMRI investigations. In addition, GSDs provide a basis for understanding the general problem of object perception and reasoning.

Acknowledgements. This chapter represents an updating of Biederman (2000). This research was supported by NSF 0420794, 0426415, and 0531177.

References

Baker C, Behrmann M, Olson C (2002) Influence of visual discrimination training of the representation of parts and whole in monkey inferotemporal cortex. Nat Neurosci 5:1210–1216

Behrmann M, Peterson M, Moscovitch M, Suzuki S (2006) Independent representation of parts and the relations between them: evidence from integrative agnosia. J Exp Psychol Hum Percept Perform (32:1169–1184)

Biederman I (1987) Recognition-by-components: a theory of human image understanding. Psychol Rev 94:115–147

Biederman I (1995) Visual object recognition. In: Kosslyn SM, Osherson DN (Eds) An invitation to cognitive science, 2nd edition, Vol. 2, visual cognition. MIT Press, Cambridge MA, pp 121–165

Biederman I (2000) Recognizing depth-rotated objects: a review of recent research and theory. Spat Vis 13:241–253

Biederman I, Bar M (1998) Same-different matching of depth-rotated objects. Paper presented at the Meetings of the Association for Research in Vision and Ophthalmology, Ft. Lauderdale, FL., May. Invest Ophthalmol Vis Sci 39:1113

Biederman I, Bar M (1999) One-shot viewpoint invariance in matching novel objects. Vision Res 39:2885–2899

Biederman I, Bar M (2000) Views on views: response to Hayward & Tarr (2000). Vision Res 40:3901–3905.

Biederman I, Cooper EE (1991) Evidence for complete translational and reflectional invariance in visual object priming. Perception 20:585–593

Biederman I, Cooper EE (1992) Size invariance in visual object priming. J Exp Psychol Hum Percept Perform 18:121–133

Biederman I, Gerhardstein PC (1993) Recognizing depth-rotated objects: evidence and conditions for 3D viewpoint invariance. J Exp Psychol Hum Percept Perform 19:1162–1182

Biederman I, Gerhardstein PC (1995) Viewpoint-dependent mechanisms in visual object recognition: reply to Tarr and Bülthoff (1995). J Exp Psychol Hum Percept Perform 21:1506–1514

Biederman I, Kalocsai P (1997) Neurocomputational bases of object and face recognition. Philos Trans R Soc London Biol Sci 352:1203–1219

Biederman I, Subramaniam S (1997) Predicting the shape similarity of objects without distinguishing viewpoint invariant properties (VIPs) or parts. Invest Ophthalmol Vis Sci 38:998

Biederman I, Subramaniam S, Bar M, Kalocsai P, Fiser J (1999) Subordinate-level object classification reexamined. Psychol Res (62:131–153)

Cooper EE, Biederman I, Hummel JE (1992) Metric invariance in object recognition: a review and further evidence. Can J Psychol 46:191–214

Cooper EE, Subramaniam S, Biederman I (1995) Recognizing objects with an irregular part. Invest Ophthalmol Vis Sci 36:473

Edelman S, Bülthoff HH (1992) Orientation dependence in the recognition of familiar and novel views of 3D objects. Vision Res 32:2385–4000

Esteky H, Tanaka K (1998) Effects of changes in aspect ratio of stimulus shape on responses of cells in the monkey inferotemporal cortex. Soc Neurosci Abstr 24:899

Hayward WG, Tarr MJ (1997) Testing conditions for viewpoint invariance in object recognition. J Exp Psychol Hum Percept Perform 23:1511–1521

Hayworth KJ, Biederman I (2005) Differential EMRI activity produced by variation in parts and relations during object perception. J Vision 5(8):740.

Hummel JE, Biederman I (1992) Dynamic binding in a neural network for shape recognition. Psycholog Rev 99:480–517

Humphreys GW, Riddoch MJ (1987) Visual object processing: a cognitive neuropsychological approach. Lawrence Erlbaum Associates Ltd, London

Kayaert G, Biederman I, Vogels R (2003) Shape tuning in macaque inferior temporal cortex. J Neurosci 23:3016–3027

Kayaert G, Biederman I, Op De Beeck H, Vogels R (2005b) Tuning for shape dimensions in macaque inferior temporal cortex. Eur J Neurosci 22:212–224

Kayaert G, Biederman I, Vogels R (2005a) Representation of regular and irregular shapes in macaque inferotemporal cortex. Cereb Cortex 15:1308–1321

Kobatake E, Tanaka K (1994) Neuronal selectivities to complex object features in the ventral visual pathway of the macaque cerebral cortex. J Neurosci 71:856–867

Kourtzi Z, Kanwisher N (2000) Cortical regions involved in perceiving object shape. J Neurosci 20:3310–3318

Lades M, Vortbrüggen JC, Buhmann J, Lange J, von der Malsburg C, Würtz RP, Konen W (1993) Distortion invariant object recognition in the dynamic link architecture. IEEE Trans Comput 42:300–311

Logothetis NK (1999) Paper presented at a workshop on Visual Object Recognition by Humans and Machines, Bad Homburg, Germany, May.

Logothetis NK, Pauls J, Bülthoff HH, Poggio T (1994) View-dependent object recognition by monkeys. Curr Biol 4:401–414

Marr, D. (1982). *Vision*. San Francisco: W.H. Freeman.

Mel BW (1997) SEEMORE: combining color, shape, and texture histogramming in a neurally-inspired approach to visual object recognition. Neural Comput 9:777–804

Nederhouser M, Biederman I, Davidoff J, Yue X, Kayaert G, Vogels R (2005) The representation of shape in individuals from a culture with limited contact with regular, simple artifacts. Paper presented at the 5th Annual Meeting of the Vision Sciences Society. May.

Nowak LG, Bullier J (1997) The timing of information transfer in the visual system. In: Kaas J, Rockland K, Peters A (Eds) Extrastriate cortex, cerebral cortex Vol. 12. Plenum, New York, pp 205–241

Pasupathy A, Connor CE (1999) Response to contour features in macaque V4. J Neurophysiol 82:2490–2502

Poggio T, Edelman S (1990) A network that learns to recognize three-dimensional objects. Nature 343:263–266

Riesenhuber M, Poggio T (2002) Neural mechanisms of object recognition. Curr Opin Neurobiol 12:162–168

Shepard RN, Cermak GW (1973) Perceptual-cognitive explorations of a toroidal set of free-form stimuli. Cognit Psychol 4:351–377

Tanaka K (1993) Neuronal mechanisms of object recognition. Science 262:685–688

Tanaka K (1996) Inferotemporal cortex and object vision. Annu Rev Neurosci 19:109–139

Tarr MJ, Bülthoff HH (1995) Is human object recognition better described be geon structural descriptions or by multiple views? Comment on Biederman and Gerhardstein (1995). J Exp Psychol Hum Percept Perform 21:1494–1505

Tarr MJ, Bülthoff HH, Zabinski M, Blanz V (1997) To what extent do unique parts influence recognition across viewpoint? Psychol Sci 8:282–289

Tarr MJ, Williams P, Hayward WG, Gauthier I (1998) Three-dimensional object recognition is viewpoint dependent. Nat Neurosci 1:275–277

Tsunoda K, Yamane Y, Nishizaki M, Tanifuji M (2001) Complex objects are represented in macaque inferotemporal cortex by the combination of feature columns. Nat Neurosci 4:832–838

Tversky A (1977) Features of similarity. Psychol Rev 84:327–352

Tversky B, Hemenway K (1984) Objects, parts, and categories. J Exp Psychol Gen 113:169–193

Vogels R, Biederman I (2002) Effects of illumination intensity and direction on object coding in macaque inferior temporal cortex. Cereb Cortex 12:756–766

Vogels R, Biederman I, Bar M, Lorincz A (2001) Sensitivity of macaque temporal neurons to differences in view-invariant *vs* metric properties of depth-rotated objects. J Cogn Neurosci 13:444–453

Winston PH (1975) Learning structural descriptions from examples. In: Winston PH (Ed) The psychology of computer vision. McGraw-Hill, New York, pp 157–209

Wiscott L, Fellous J-M, Krüger N, von der Malsburg C (1997) Face recognition by elastic bunch graph matching. IEEE PAMI 19:775–779

Zerroug M, Nevatia R (1996) Volumetric descriptions from a single intensity image. Int J Comput Vis 20:11–42

6
Object Recognition in Humans and Machines

Christian Wallraven and Heinrich H. Bülthoff

1 Introduction

The question of how humans learn, represent and recognize objects has been one of the core questions in cognitive research. With the advent of the field of computer vision – most notably through the seminal work of David Marr – it seemed that the solution lay in a three-dimensional (3D) reconstruction of the environment (Marr 1982, see also one of the first computer vision systems built by Roberts et al. 1965). The success of this approach, however, was limited both in terms of explaining experimental results emerging from cognitive research as well as in enabling computer systems to recognize objects with a performance similar to humans.

More specifically, psychophysical experiments in the early 1990s showed that human recognition could be better explained in terms of a view-based account, in which object representations consist of snapshot-like views (Bülthoff and Edelman 1992) instead of a full, 3D reconstruction of the object. The most important result of these experiments is that recognition performance is critically dependent on the amount of view-change between learned and tested object view. This stands in stark contrast to the predictions from frameworks using 3D representations such as the often-cited Recognition-By-Components theory (Biederman 1987) which is based on a 3D alphabet of basic geometrical shapes (so-called geons) and predicts a largely view-invariant recognition performance. To date, psychophysical and neurophysiological experiments have provided further evidence for the plausibility of the view-based approach (see, e.g., Tarr and Bülthoff 1998; Wallis and Bülthoff 2001 for two recent reviews).

Max Planck Institute for Biological Cybernetics, Spemannstrasse 38, Tübingen, Germany

In a recent paper, an attempt has been made to reconcile these two approaches to object processing (Foster and Gilson 2002): a careful study of view-dependency of novel objects that were created by combining structural properties (number of parts) with metric properties (thickness, size of parts) has found that both view-dependent and view-independent processing seem to be combined in object recognition. Thus, instead of taking the extreme perspective of either view-based or view-invariant processing one might envisage a visual processing framework in which features are selected according to the current task, where the optimality, efficiency and thus the dependency on viewing parameters of the features depend on the amount of visual experience with this particular task.

Robust extraction of structural, view-invariant features from images, however, has proved to be difficult for computer vision. Therefore, parallel to view-based approaches to object recognition in human psychophysics, view-based computer vision systems began to be developed. These sometimes surprisingly simple recognition systems were based on two-dimensional representations such as histograms of pixel values (Swain and Ballard 1991), local feature detectors (Schmid and Mohr 1997) or on pixel representations of images (Kirby and Sirovich 1990). The good performance of these recognition systems using complex images taken under natural viewing conditions can be seen as another indicator for the feasibility of a view-based approach to recognition.

To date, most theories of object recognition as well as most computer vision systems have mainly focused on the *static* domain of object recognition. Visual input on the retina, however, consists of dynamic changes due to object- and self-motion, non-rigid deformations of objects, articulated object motion as well as scene changes such as variations in lighting, occluding and re- and disappearing objects, and at any given point in time several of these changes can be interacting. The central question for this chapter will thus be: To what extent do object recognition processes rely on *dynamic* information per se? Several psychophysical experiments, which will be discussed below, suggest an important role for dynamic information, in both learning and recognition of objects. Based on these findings, an extension of the current object recognition framework is needed in order to arrive at truly spatio-temporal object representations.

In this chapter, we therefore want to explore the idea of learning and representing objects in a spatio-temporal context by developing a computational object recognition framework motivated by psychophysical results. Specifically, we are interested in developing a recognition framework, which can learn and recognize objects from natural visual input in a continuous perception-action cycle. In the following, we will first briefly summarize the psychophysical experiments that guided the development of the recognition framework. Subsequently, we will present details of the framework together with results from several computational recognition experiments. Finally, we will summarize experiments conducted with a humanoid robot in which the framework was applied to multi-modal recognition of objects using proprioceptive and visual input. These experiments represent a first step towards a closely coupled perception-action system based on and motivated by psychophysical research.

2 Psychophysical Experiments

2.1 Temporal Continuity for Object Learning

To illustrate the importance of temporal context, consider a view-based object recognition system faced with the task of learning object representations. Input to this system consists of a series of views that the system acquires. The problem for this recognition system is how to link the many views of an object to create a consistent and coherent object entity; especially since these views can be very different from each other. One solution to this problem is the observation that in real life we seldom see only isolated snapshots of objects. Usually novel objects are explored either actively through manipulation by our hands or by walking around them. This results in a sequence of images that gradually change from the initial view of the object to a very different one within a short period of time – temporal contiguity. This general observation about natural visual input in a continuous perception-action context motivates the following question: Does the human visual system use temporal contiguity to build a mental representation of the object in order to associate views together? This temporal association hypothesis was investigated in two studies (Wallis and Bülthoff 2001; Wallis 2002), which we briefly review below.

Study 1 – Stimuli: Twelve faces from 3D-laser-scanned female heads were used as stimuli. The faces were separated into three training sets of four faces each. Using a technique by Blanz and Vetter (1999), 3D morph sequences between all possible combinations of face pairs within each set were created. A sequence consisted of a left profile view (–90°) of an original face A, a –45° view of morph A→B (the average of face A and B), a frontal view (0°) of face B, a +45° view of morph A→B, and finally a right profile (+90°) of face A (Fig. 1). A backward

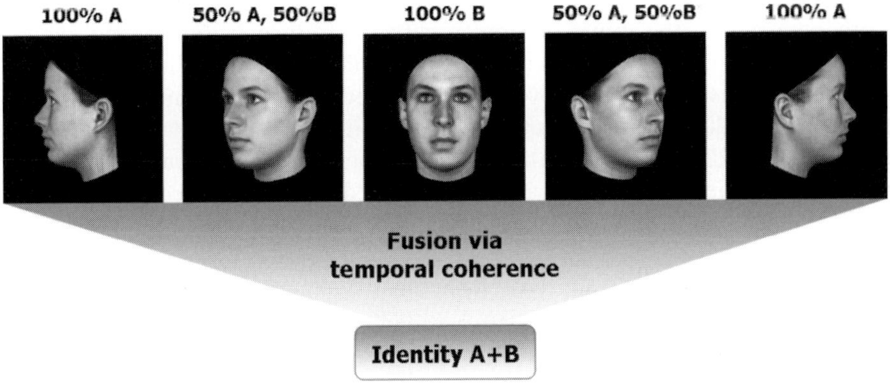

FIG. 1. Illustration of the morph experiment. A morph sequence of two individuals (A and B) is shown to participants who fuse the sequence into one coherent identity due to the temporal continuity present in the visual input

sequence showed the same images in reversed order. The training sequence consisted of a forward sequence and a backward sequence, followed by the forward sequence again and the backward sequence again.

Study 1 – Experimental design: Participants were divided into two groups. In the first group, each participant was trained using sequential presentation of the sequences. In the second group, training used simultaneous presentation of all morph images shown together on the computer screen for the same total time. After training, the participants performed a simple image matching task in which they had to decide whether two subsequently shown images were different views of the same face or not. Half of the trials presented matches, whereas in the other trials half of the test face pairs belonged to the same training set (within set, WS) and the other half to different training sets (between set, BS). If views of objects are associated based on temporal contiguity, then training with sequential presentation should cause the images grouped in one training sequence to be fused together as views of a single object. After such training, participants in the testing phase would be expected to confuse faces that were linked together in a training sequence (WS) more often than between-faces that were not (BS). Training with simultaneous presentation was included to rule out the possibility that the morphs alone were sufficient for the training effect, in which case an effect should appear after both training procedures.

Study 1 – Results: The results of the experiment confirmed that participants were more likely to confuse those faces that had been associated temporally in a sequence (WS). Thus, participants learned to fuse arbitrary views of different faces into one coherent identity without any explicit training. In addition, the results from the second group indicated that the presence of morphs among the training images alone was not sufficient to cause the association of two different faces with each other.

Study 2 – Stimuli and design: In a follow-up study (see Wallis 2002), the results were replicated using two different sets of stimuli for sequential presentation. The sequences here consisted of images of *different* faces instead of morphs thus further increasing the visual difference between frames. In the second experiment, training sequences were created by scrambling the poses in the sequence such that at most two consecutive images showed a consistent and smooth rotation (of 45°). The remaining experimental parameters in the two experiments closely followed the design of the first study for the morph group. This experiment tested whether temporal association based on temporal contiguity could still be detected even when the spatial similarity between consecutive images was low.

Study 2 – Results: The main result from the first experiment was that confusion scores in the WS condition were significantly higher than those in the BS condition indicating that temporal association, indeed, is possible even with more dissimilar sequences. However, the relative effects of temporal association on the two test conditions were *reduced* compared to that of the morphed sequences in the first study. This is an important finding as it indicates that association is influenced by *spatial similarity as well as temporal contiguity*. In the second experiment, there were no significant main effects for a scrambled presentation of images, which destroyed the consistent rotation interpretation but otherwise should have left the pure temporal contiguity intact. However, over the course of three blocks, a significant trend towards a slow dissociation between the two test conditions could be detected. The author interpreted this as a sign that temporal association can take place under such conditions – albeit at a much slower rate.

2.2 General Discussion

Summarizing the two studies, one can conclude that the learning of object representations is strongly influenced by the temporal properties of the visual input. One successful strategy of how the brain might solve the task of building consistent object representations – even under considerable changes in viewing condition – seems to be to assign consecutive images to the same object. This process is not only influenced by temporal parameters but also to a significant degree by the similarity properties of the input. Arbitrary images seem to be much harder to learn, suggesting a crucial influence of the spatial similarity of visual input. These findings therefore are consistent with the extended concept of *spatio-temporal continuity* resulting in integration of images that are below a certain similarity threshold and that are presented within a certain time window.

The findings of these experiments as well as further psychophysical (most notably Stone 1999) and physiological studies (e.g., Miyashita 1988) provide strong evidence for an integral role of temporal characteristics of visual input in object representations and for their active use in learning and recognizing objects. However, the question remains how exactly spatial and temporal information can be integrated in object representations. In the next chapter, we propose a computational implementation, which provides such an integration as part of the recognition and learning procedure.

3 Computational Recognition System

3.1 The Keyframe Framework

The abstract framework shown in Figure 2 consists of several key elements. First, and most importantly, the system processes incoming images in a sequential manner in order to extract so-called keyframes, which represent an extension of the view-concept followed in the view-based approach. Each input frame of an image sequence is first processed in order to extract local features (so-called interest points), which are then tracked across subsequent frames. Eventually, the changes in visual input will be too large and will lead to a loss of tracked features. The core idea behind the framework is that keyframes are precisely defined by that point at which tracking breaks down. If this happens, a new keyframe is inserted into the object representation and the process repeats.

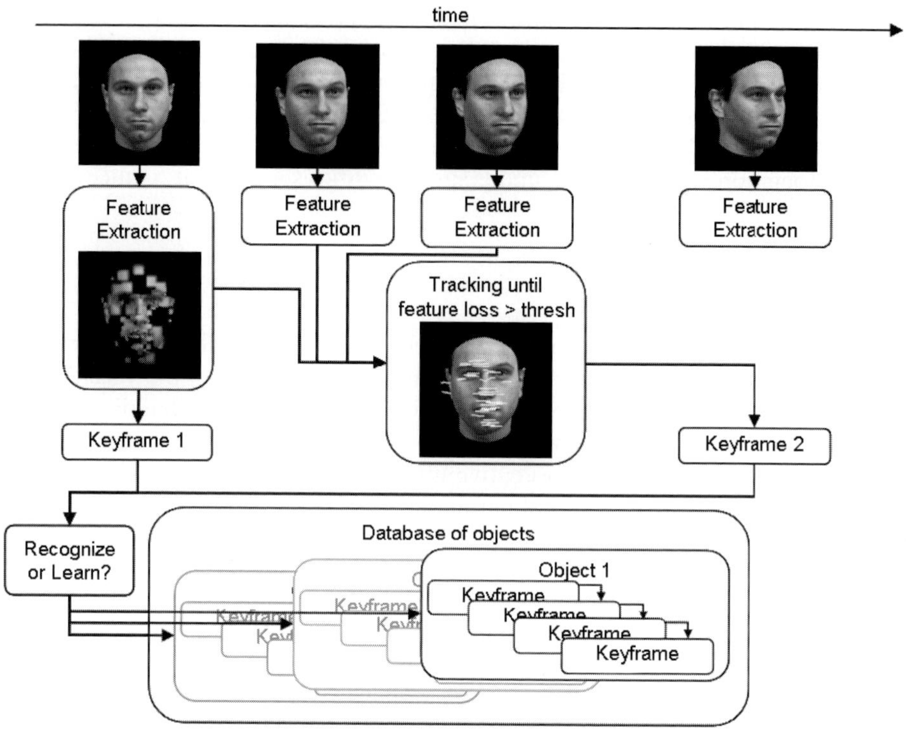

FIG. 2. Abstract description of the keyframe framework. Local feature tracking is used to extract visual events ("keyframes") from an image sequence, resulting in a view-based, connected object representation

Keyframes are thus two-dimensional views (snapshots) of the scene, which are defined by the temporal continuity (in close connection to the psychophysical experiments described in the previous section) of the visual input and form a connected graph of views (see Fig. 2).

Note that in this abstract form the keyframe approach resembles the concept of "aspect graphs" (Koenderink and van Doorn 1979), in which objects are defined by their aspects, i.e., by visual events, where a sudden change in the observed shape of the object occurs. Even though the rigorous mathematical formulations were highly appealing to the computer vision community due to their geometric interpretations, computational realizations of the aspect graph framework for arbitrary objects proved to be difficult. One of the core ideas, however, namely the representation of objects by visual events remains a powerful concept, which our proposed framework retains. Whereas the focus of aspect graphs mainly lies in representations of 3D objects by well-defined views, we want to go one step further with the keyframe concept by representing all kinds of dynamic visual input with the help of two-dimensional views.

Furthermore, learning and recognition are not separated in our framework with new keyframes constantly being compared against the learned library. This means that the system continuously learns new data that can be used to augment existing object representations or form new ones. This is a crucial pre-requisite for any cognitive system, as it is embedded in a dynamic sensory environment and thus constantly receives new input that has to be evaluated and categorized in order to create appropriate (re-)actions.

This embedding of object learning and recognition in a temporal context is reminiscent of the "active vision" paradigm that was developed in the 1980s in computer vision (for example, Aloimonos et al. 1987). Most of the research in active vision was focused on how to control the optics and mechanical structure of vision sensors to simplify the processing for computer vision. Here, we go one step further by endowing object representations themselves with a temporal component through tracking of features and the graph-like keyframe representation.

3.2 Properties of the Framework

As indicated in the introduction, learning and recognition of objects certainly seems possible using only the static dimension – one of the key questions then of course becomes: What – apart from psychophysical motivations – is the advantage of using the temporal dimension in the framework?

Keyframes: In the most extreme case of a view-based framework, learning would involve storing all input images. This strategy is certainly not feasible for any reasonable amount of learning data due to storage constraints. In addition, it also represents a severe problem for recognition as the time it takes to index into the representation becomes prohibitively large. The question thus is: which views to select for learning? Here the keyframe concept provides an intuitive answer to that question: select the views in which an important visual event occurs. In order for this strategy to be successful, one needs to make the assumption that the visual input is, on average, slowly changing, which, given the psychophysical evidence presented above, certainly seems to be valid. Furthermore, the keyframes are organized in a directed graph structure, which allows for pre-activation of frames during recognition of image sequences. If two connected keyframes in a row could be recognized, chances are good that the next incoming keyframe will be the next node in the graph. This strategy thus dramatically reduces the search time during recognition of known sequences or sequence-parts.

Visual features: We chose to include local features in the framework in order to focus on locally informative visual aspects of each frame (see Fig. 2). These local features consist of simple image fragments (much in the spirit of Ullman et al. 2002) extracted around interest points that are detected in the image at several scales. Whereas of course the exact nature of these features is open to further experimentation (for example, Krieger et al. 2000; Lowe 2004 for other approaches), already these relatively simple image fragments are effective in compressing image data. In addition, the most important contribution of the tracking that is used to determine the keyframes is that it allows access to feature trajectories. In our framework, the trajectories follow features from one keyframe to the next. The larger the visual difference between keyframes, the more discriminative these feature trajectories are – this is because the chances of false matches are reduced, the longer a feature can reliably be tracked (see also Tomasi and Kanade 1991). More importantly, the trajectories describe the transformation of each feature from one keyframe to another and thus can be used to generate priors for matching feature sets. Consider, for example, a sequence of a rotating object for which the feature trajectories between keyframes will have a shape that is specified by the direction of the (3D) rotation. For recognition, a matching prior can now be derived directly from the trajectories by constraining feature matches to that very direction. Whereas this strategy obviously works only for some simpler cases of object motion, it nevertheless will provide a much more robust feature matching. In addition, we want to stress that our focus on visual features and their transformations between visual events is a much broader concept not restricted to object motion alone. Going beyond a simple matching prior, this information can also be used to explicitly model generic object or category transformations (Graf 2002), which expands the keyframe framework into a general learning concept for any dynamic visual data.

3.3 Computational Experiments

In the following, we will briefly present results from computational experiments, in which we tested the performance of the keyframe implementation on a highly controlled database (for details of the implementation as well as additional experiments also including real-world video sequences, see Wallraven and Bülthoff 2001).

Stimuli: The database consisted of 60 sequences of faces taken from the MPI face-database (Troje and Bülthoff 1996). This database contains highly realistic 3D laser-scans of faces and allows full control of all aspects of rendering (pose, lighting, shadows, scene, etc.) for benchmarking recognition algorithms. Each sequence showed a face turning from –90° (left) profile view to +90° (right) profile view consisting of 61 frames at pose intervals of 3 degrees. All faces were rendered from a viewing distance of 1.3 m on a black background using a frontal point-light source. Our test sets consisted of images from the same sequences in addition to novel images containing pose variations of +/–15° (upwards and downwards) as well as two different illumination directions.

Keyframes: Using the local feature tracking algorithm described above, the system found 7 keyframes for each of the 60 sequences (Fig. 3a shows some example keyframes and their average poses). Furthermore, the angular distance between subsequent keyframes is smallest for the frontal poses (between keyframes 3 and 5). This is due to the fact that a rotation around the frontal view causes larger variations in features (such as ears disappearing and appearing) leading to an earlier termination of tracking. Note also that even

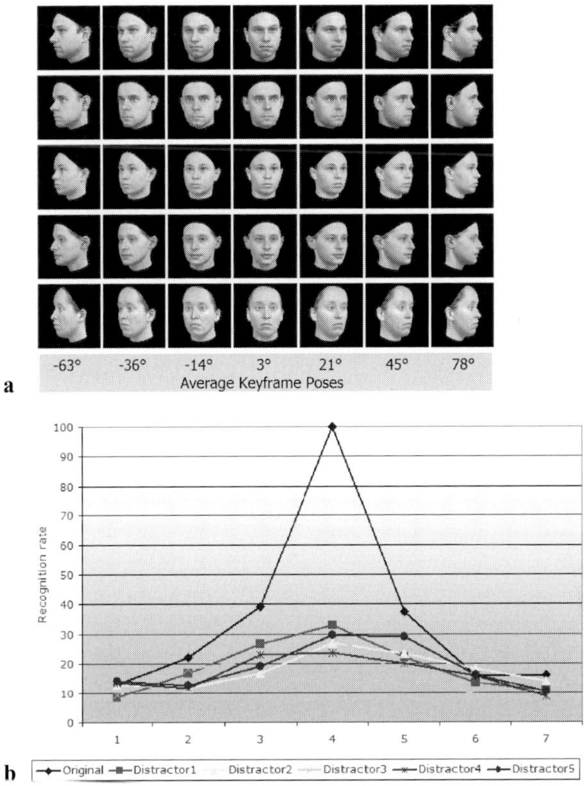

FIG. 3. **a** Examples of keyframes extracted from image sequences of rotating faces. The bottom figures list the average pose in degrees found across the whole database. **b** Matching scores for 6 "novel" faces. Note that the target face has a much higher matching score than the five other distractors

though the framework itself has no prior knowledge about the object class or the object motion, similar views are selected as keyframes. This is a demonstration that our framework is able to generate consistent representations provided the input also exhibits consistent characteristics. Finally, the representation of each image sequence consists of a number of keyframes containing local features, resulting in a significant, size reduction. This is an essential property for any view-based system working with dynamic data since otherwise huge amounts of data would have to be stored. To calculate the size reduction of the representation, we compared the size of the final sequence models to the raw pixel data and determined a reduction of 99.1% (7 keyframes compared to 61 original frames corresponds to a reduction of ~90%; each keyframe contains ~200 local features, each of which consists of 5×5 pixels. Given the original image size of 256×256 pixels, this results in a reduction of ~92% per keyframe).

Recognition results: Our first recognition experiment concerned a validation of whether the resulting keyframe representation could support recognition of intermediate views of the original sequences. We therefore tested all keyframe representations with the remaining $(61 - 7)*30 = 1620$ frames not included in the keyframe representation, which resulted in a total recognition rate of 100%. To illustrate the robust matching, Figure 3b shows the matching scores for a set of keyframes with one target image and 5 distractor images, all of which show the same view. First of all, one can see that the target has a much higher matching score than the distractors. Interestingly, the highest match score for the distractors is almost exclusively achieved for the correct pose. In addition, all curves show a consistent view-based behaviour with a fall-off around the best matching keyframe. Recognition rates in the second experiment testing novel views under pose and illumination variation were 98.6% and 89.4%, respectively. Although pose variation is tolerated well by the system, changes in illumination clearly show the limits of the simple matching scheme. Taking the amount of compression into account, however, we think that these results demonstrate the feasibility and robustness of our approach (see also Wallraven and Bülthoff 2001).

4 Multi-Modal Keyframes

So far, the keyframe framework has been treated as a framework for recognition of objects in the visual modality. The general idea of spatio-temporal object representations, however, can of course be extended to other modalities as well. In the following, we will introduce such a multi-modal object representation combining visual with proprioceptive information, which was successfully implemented on a robot-setup and subsequently tested in object learning and recognition scenarios.

Recent research in neuroscience has led to a paradigm shift from cleanly separable processing streams for each modality towards a more integrative picture consisting of multi-modal object representations. Such cross-modal integration of data from different modalities was also shown, for example, to play an important role for haptic and visual modalities during object recognition. In a recent psychophysical experiment (see Newell et al. 2001), participants had to learn views of four simple, 3D objects made of stacked LEGO™ bricks either through the haptic modality (when they were blind-folded) or through the visual modality (without being able to touch the objects). Testing was then done using an old-new recognition paradigm with four different conditions: two within-modality conditions, in which participants were trained and tested in either the haptic or the visual domain and two between-modality conditions, in which information from the learned modality had to be transferred to other modalities in order to solve the recognition task. For each condition, in addition, either the same viewpoint or a viewpoint rotated 180° around the vertical axis was presented in order to test the viewpoint-dependence of object recognition.

The recognition results for the four conditions showed first of all that cross-modal recognition occurred at levels well above chance. Not surprisingly, recognition of rotated objects in the within-modality condition was severely affected by rotation in both modalities. This shows that not only visual recognition is highly view-dependent but also that haptic recognition performance is directly affected by different viewing parameters. One could thus extend the concept of view-based representations of objects also to the haptic modality. Another interesting finding of this study is that recognition performance in the haptic-to-visual condition increased with rotation. The authors assumed that this was an example of a true cross-modal transfer effect – the reason for such a transfer lies in the fact that during learning the haptic information extracted by participants was mainly derived from the back of the object. When presented with a rotated object in the visual modality, this haptic information was now visible, which enabled easier recognition. The results from this experiment thus support the view that haptic recognition is also mediated by view-based processes – although the exact dependence on viewing angle remains to be investigated. In addition, the authors shed light on how information from the haptic modality can be used to enable easier recognition in the visual modality. Taken together with the spatio-temporal framework outlined above, this cross-modal transfer might be an important reason for the excellent visual performance of human object recognition – after all, it is known that infants learn extensively by grasping and touching objects, which thus could provide a "database" of object representations for visual recognition.

4.1 Multi-Modal Keyframes – the View-Transition Map

Taking these psychophysical experiments as inspiration, we now want to describe how visual and proprioceptive input can be combined to create and test a multi-modal keyframe representation.[1]

Let us consider a person who is examining an object by holding it in their hand and turning it around – the sensory information that is available in this situation consists of not only dynamic visual data but also dynamic haptic information. More specifically, we will focus on the proprioceptive information as a subset of the haptic information, which consists of the 3D configuration of the hand (such as the exact configuration of the fingers holding the object) as well as that of the wrist. How could this information be of use for learning and recognition?

First of all, proprioceptive information about the 3D configuration of the hand could actually be used in a similar manner as in the psychophysical experiment described in the previous section. Since it is three-dimensional, it can for example generate a 3D viewing space in which keyframes (derived from the visual infor-

[1] The multi-modal representation, as well as the experiments were developed in collaboration with Sajit Rao, Lorenzo Natale, and Giulio Sandini at the Dipartimento di Informatica, Sistemistica e Telematica at the University of Genoa.

mation of the image sequence) can be anchored at proprioceptive coordinates. This would link the visual appearance from the keyframe with the hand position and configuration and thus provide a proprioceptively anchored visual space. Returning to Figure 2, we see that one of the inherent disadvantages of the keyframe framework is that the real-world topology of the keyframe graph is undefined – only the outgoing and incoming links for each keyframe are known. Although this provides enough information to resolve recognition tasks (see previous section), being able to convert the viewer-centered keyframe graph into an object-centred keyframe graph would provide additional constraints for matching visual appearances since such a representation would be more closely integrated into a perception-action loop.

One of the problems with the idea of proprioceptive space, however, is that absolute coordinates in such a space make little sense from the perspective of recognition. Although it might be the case that objects suggest a canonical grasp (in much the same manner as they might suggest an affordance in the Gibsonian sense), usually it is possible to pick up and hold an object in a number of ways – all of which will change the absolute proprioceptive coordinates to which keyframes will be attached. Our solution is to interpret the proprioceptive space in a similar manner as the keyframe graph: as a representation based on *changes* in its underlying modality. Thus, rather than using an absolute frame of reference, each generated keyframe could be attached to a relative change in proprioceptive coordinates. One way to implement such a multi-modal representation is as a lookup table, in which each entry can be accessed via its relative change in proprioceptive space – this change can, for example, be simply the difference between the proprioceptive state vectors of the hand (including wrist angles, finger positions, etc.). This novel representation – which we call a view transition map – would for n visual keyframes consist of $n(n-1)$ entries for all possible proprioceptive transitions between keyframes.

How could one use this view-transition map to recognize objects? First of all, a keyframe representation of an object is learned in an active exploration stage using a pre-learned motor program, which for example grasps an object and turns it around. Each new keyframe is entered into the transition map at a position specified by the relative change in proprioceptive state from the previous keyframe. In addition, the transition map is enlarged by adding transitions from this keyframe to all previous ones. In a second step, a test object is picked up and keyframes are extracted again while the same motor program is executed. In order to recognize this object using the transition map, the first keyframe that was generated from the test sequence is matched against all of the keyframes of the training sequence using visual similarity (in our implementation, similarity consisted of simple Euclidean distance – using local feature matching would further increase the robustness of the system). Once this match has been established, the transition map can be used to quickly find neighboring keyframes by looking for the most similar proprioceptive transition from the keyframe that matches the current change in the proprioceptive state. With this strategy one could expect to recognize objects in a much more efficient manner as indexing proprioceptive transitions allows for direct matches in an object-centered reference frame.

4.2 Computational Recognition Experiment

The proposed view transition map representation was tested in a computational experiment in which we explored its use for object recognition.

Experimental setup: Figure 4a shows the robot setup from the Dipartimento di Informatica, Sistemistica e Telematica at the University of Genoa that was

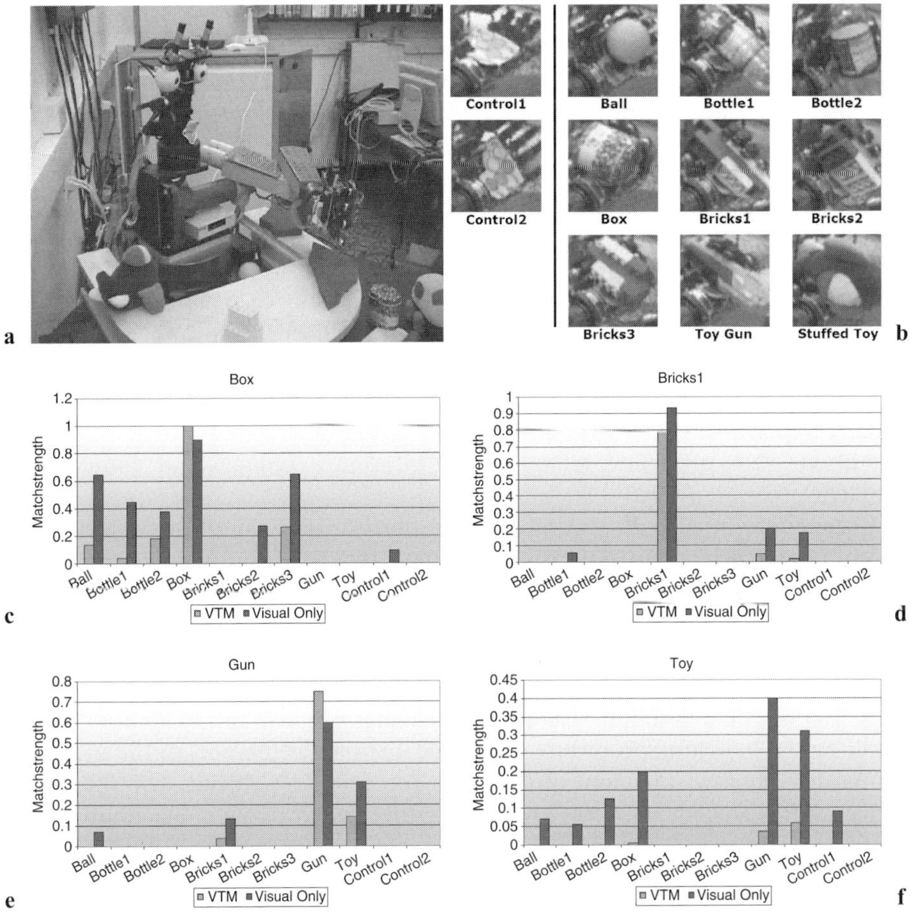

FIG. 4. **a** The robot setup (Metta et al. 2000) that was used in the multi-modal keyframe experiments **b** the objects used in the learning and recognition experiments. **c–f** Example results from the object recognition experiment showing the increase in discriminability when using multi-modal representations. The bright bars show matching using the view transition map, the dark bars show visual matching only

used in this experiment (Metta et al. 2000). The important components of the setup for this experiment consist of an actively foveating stereo camera head (using space-variant image sensors mimicking the human retinal structure) and an anthropomorphic robotic arm with a fully controllable hand. The camera head was pre-programmed to fixate on the location of the hand in order to track the hand during all movements. In addition, a trajectory for the hand movement was defined, which consisted of the hand rotating first around the axis defined by the arm ("turning the hand") and then around a second axis resulting in an up-and-down movement of the hand. This exploratory motion sequence ensured an adequate visual coverage of any grasped object. The test objects for the experiments consisted of 9 household and toy objects and are depicted in Figure 4b – note that some of the objects are rather similar in terms of their visual appearance.

In order for the robot to learn an object, it was placed into the robot's hand and the exploratory motion sequence was initiated. The visual input from the foveated cameras was then used to track local features in real-time using the keyframe framework as described in the previous section. Each time the system found a keyframe, the proprioceptive transition leading from the last to the current keyframe was used as an index into a matrix where each entry stored the visual information of the frame (in this case simply consisting of the whole frame rather than its local feature representation). In addition, each incoming keyframe was matched against all existing keyframes using the view-transition map matching procedure outlined above. If a match of suitable strength was found, the keyframe was discarded, otherwise the keyframe was inserted into the representation. A total of 9 objects were learned in this manner; in addition two control conditions were recorded, which simply showed a sequence of empty hand moving.

Recognition results: To test recognition performance, six of the objects were given again to the robot and the same movements were executed. Each new keyframe was then compared against all learned transition maps using the matching procedure described above and the amount of matches in each transition map was added up to a final matching score. If the sequence would be identical, all keyframes would be found in the map and therefore the matching score would be 1. To provide a baseline, visual-only matching was also run in addition to the multi-modal matching procedure. Figure 4c–f shows histograms of the matching scores for the two matching procedures for four test-objects. For the "box" object, both procedures correctly predict the right category; the multi-modal matching, however, has a much higher discriminability compared to the visual-only matching. The same is true for the "bricks1" and "gun" object. Finally, the "toy" object is an example of an object, which is not correctly recognized by visual-only but is recognized by the multi-modal matching.

Summary: The results of these initial computational experiments are very encouraging. Through a straightforward extension of the keyframe approach to include proprioceptive information, we have shown how multi-modal object representations can be learned as well as how such representations can help to increase the discriminability of object recognition. Since our representation is in part three-dimensional (i.e., coupled to proprioceptive coordinates), some of the robustness actually comes from 3D information in a viewer-/manipulator-centred coordinate system. It would be interesting to see how such a representation might capture the results in the chapter by Gschwind et al. (this volume) on exploration of 3D shapes.

Among several extensions that can be envisioned, adding more sophisticated local feature matching, better classification schemes as well as different cue combination approaches should further improve the performance of the framework. Another interesting property of the transition map is that it enables execution of specific motor actions based on visual input. Consider, for example, a situation in which an object has to be manipulated in order to insert it into a slot. The inverse of the transition map would allow such a task to be solved by executing motor commands that trace out a valid motor path to the desired view based on the current view. In a similar manner, the transition map could also be used for efficient imitation learning based on visual input and for executing mental rotations. The key to all of these applications is that the transition map provides a strong coupling between proprioceptive data (action) and visual data (perception) and in this manner facilitates representation of a perception-action loop in an effective and efficient way.

5 Conclusion

In this chapter, we proposed an abstract framework for learning and recognition of objects that is inspired by recent psychophysical results which have shown that object representations in the human brain are inherently spatio-temporal. In addition, we have also presented results from a computational implementation of this keyframe framework, which demonstrate that such a system can reliably recognize objects under a variety of conditions. Finally, experiments with multimodal keyframes have shown that by integrating non-visual cues, object learning and recognition becomes more efficient and effective. We believe that this framework can represent a significant step in designing and implementing a truly cognitive system, which is embedded in a constantly changing environment and thus has to constantly analyze and learn in order to plan its (re-)actions.

References

Aloimonos JY, Weiss I, Bandopadhay A (1987) Active vision. Int J Comput Vis 1:333–356

Biederman I (1987) Recognition-by-components: a theory of human image understanding. Psychol Rev 94:115–147

Blanz V, Vetter T (1999) A morphable model for the synthesis of 3d faces. Proc ACM SIGGRAPH 1999:187–194

Bülthoff HH, Edelman S (1992) Psychophysical support for a 2-d view interpolation theory of object recognition. Proc Natl Acad Sci U S A 89:60–64
Foster DH, Gilson SJ (2002) Recognizing novel three-dimensional objects by summing signals from parts and views. Proc R Soc Lond B 269:1939–1947
Graf M (2002) Form, space and object. Geometrical transformations in object recognition and categorization. Wissenschaftlicher Verlag Berlin, Berlin
Kirby M, Sirovich L (1990) Applications of the Karhunen-Loeve procedure for the characterization of human faces. IEEE Trans Pattern Anal Mach Intell 12:103–108
Koenderink J, van Doorn A (1979) The internal representation of solid shape with respect to vision. Biol Cybern 32:211–216
Krieger G, Rentschler I, Hauske G, Schill K, Zetzsche C (2000) Object and scene analysis by saccadic eye-movements: an investigation with higher-order statistics. Spat Vis 13:201–214
Lowe D (2004) Distinctive image features from scale-invariant keypoints. Int J Comput Vis 60:91–110
Marr D (1982) Vision. Freeman Publishers, San Francisco
Metta G, Panerai F, Sandini G (2000) Babybot: a biologically inspired developing robotic agent. Proc. Sixth International Conference on the Simulation of Adaptive Behaviors, 1–10
Miyashita Y (1988) Neuronal correlate of visual associative long-term memory in the primate temporal cortex. Nature 335:817–820
Newell FN, Ernst MO, Tjan BS, Bülthoff HH (2001) Viewpoint dependence in visual and haptic object recognition. Psychol Sci 12:37–42
Roberts L (1965) Machine perception of three-dimensional solids. In: Tippett J, Clapp L (Eds) Optical and electro-optical information processing. MIT Press, Cambridge MA, pp 159–197
Schmid C, Mohr R (1997) Local greyvalue invariants for image retrieval. IEEE Trans Pattern Anal Mach Intell 19:530–535
Stone JV (1999) Object recognition: view-specificity and motion-specificity. Vision Res 39:4032–4044
Swain M, Ballard D (1991) Color indexing. Int J Comput Vis 7:11–32
Tarr M, Bülthoff HH (1998) Object recognition in man, monkey, and machine. MIT Press, Cambridge MA
Tomasi C, Kanade T (1991) Detection and tracking of point features. Carnegie-Mellon Tech Report CMU-CS-91–132
Troje NF, Bülthoff HH (1996) Face recognition under varying pose: the role of texture and shape. Vision Res 36:1761–1771
Ullman S, Vidal-Naquet M, Sali E (2002) Visual features of intermediate complexity and their use in classification. Nat Neurosci 5:682–687
Wallis GM (2002) The role of object motion in forging long-term representations of objects. Vis Cogn 9:233–247
Wallis GM, Bülthoff HH (2001) Effects of temporal association on recognition memory. Proc Natl Acad Sci U S A 98:4800–4804
Wallraven C, Bülthoff HH (2001) Automatic acquisition of exemplar-based representations for recognition from image sequences. In Proc. CVPR'01 – Workshop on Models versus Exemplars

7
Prior Knowledge and Learning in 3D Object Recognition

Markus Gschwind[1], Hans Brettel[2], and Ingo Rentschler[1]

1 Introduction

Biological 3D object recognition is restricted to the sensing of 2D projections, or images, and is further constrained by the lack of transparency. The most common assumption then is that image data are referenced to mental object representations. Such representations, or object models, must be contrasted with object recognition in so far as the latter involves the understanding of image data. This distinction is central to recognition-by-components (RBC; Biederman 1987), a theory of human image understanding based on the assumption that input images are parsed into regions that display nonaccidental properties of edges. These properties provide critical constraints on the identity of 3D primitives ("geons") the images come from, e.g., cylinders, blocks, wedges, and cones, and are (relatively) invariant with viewpoint and image degradation.

RBC can be implemented by building structural representations from geons linked through explicit categorical relations (Hummel and Biederman 1992). This theory predicts that object identification will be fast and accurate if geons are readily identified in characteristic arrangements. It also implies that viewpoint invariance in 3D object recognition is achieved for all views that activate the same geon structural description (GSD; Biederman and Gerhardstein 1993). However, viewpoint invariance is not found for stimuli based on irregular blob structures ("amoebae"; Edelman and Bülthoff 1992; Bülthoff and Edelman 1992) and wire-like objects ("paper-clips"; Bülthoff and Edelman 1992). It has been held that the latter result is incompatible with recognition theories involving 3D representations. This gave rise to the multiple-views hypothesis, according to which a set of views of an object is stored in memory and the object is recognized by normalizing the input view to the most nearly compatible among such stored views (Tarr and Pinker 1989; Bülthoff and Edelman 1992).

[1]Institute of Medical Psychology, University of Munich, Goethestrasse 31, D-80336 München, Germany
[2]CNRS UMR 5141, École Nationale Supérieure des Télécommunications, Paris, France

Given these different perspectives on human object recognition, it is helpful to consider the development of object recognition by computer. Early approaches to this problem used the concept of generalized cones applied to the domain of line drawings of objects and scenes composed of polyhedral or curved parts. The understanding of such "engineering" drawings was demonstrated by producing a line drawing of the arrangement of parts as it would appear from any desired viewpoint. Yet it was clear that the interpretation of "naturalistic" images was another matter altogether (see Ballard and Brown 1982, chapter 9). To solve the latter type of problem, part-based recognition schemes are now employed in a more flexible way. For instance, the analysis of parts may be initiated by segmenting input images into regions that are recognized as parts of objects in the database. If no recognition occurs, the parameters of the initial segmentation are varied. Clearly, such approaches do not succeed in one stroke. These processes typically involve closed-loop systems where the current interpretation state is used to drive the lower level image processing functions. For these reasons, "world knowledge" and learning play key roles in second-generation image understanding and object recognition by computer (see Caelli and Bischof 1997), and the chapter by M. Jüttner, this volume).

The latter development prompted this study of the roles of prior knowledge and learning in the recognition by human observers of "structure-only" 3D objects composed of identical parts in varying spatial arrangement. As the left-right categorization of mirror-image forms is a typical feature of visual expertise (Johnson and Mervis 1997; Tanaka and Taylor 1991; Rentschler and Jüttner 2007), the test stimuli included handed objects.

2 Separating Representation and Recognition

Valid conclusions as to the nature of object representations cannot be drawn unless their dependence on stimulus information (Liu 1996; Liu et al. 1999) and task demands (Tjan and Legge 1998) is taken into account. The latter two studies made this point using an ideal observer model based on statistical pattern recognition. Thereby patterns are classified using sets of extracted features and an underlying statistical model for the generation of these patterns (see Haykin 1999).

Tjan and Legge (1998) showed that viewpoint dependence of recognition is low for structurally regular objects, but dependence increases as regularity decreases. They were further able to demonstrate a correspondence between the predicted view-point complexity (VX) of a recognition task and published human data on viewpoint dependence. For instance, they found low VX values for simple geometric objects (single geons) and mechanical compositions (distinct multiple-geon objects) consistent with the observations by Biederman and Gerhardstein (1993). By contrast, wire-like and amoebae objects showed high

VX consistent with the findings by Edelman and Bülthoff (1992). Tjan and Legge concluded that confusion about the nature of object representations can be attributed at least partly to a failure to distinguish betweeen visual processing and the type of recognition task including the physical characteristics of test objects.

The findings of Tjan and Legge would seem to be consistent with reports from object recognition by computer (see Dickinson 1993). On the one hand, 2D indexing primitives, i.e., image structures that are matched to stored object models, are useful for small object databases. The reason for this limitation is increasing search complexity and reliance on verification with decreasing complexity of primitives. On the other hand, the reliable recovery of 3D indexing primitives from input images is a very difficult problem. Nevertheless, due to a concomitant decrease in search complexity for matching, 3D indexing primitives may be more successful than 2D indexing primitives for large databases.

Against the conclusions from ideal observer models, it might be held that these models rely on traditional pattern recognition, where classification is achieved by partitioning feature space into regions associated with different pattern classes. However, there are many recognition problems that cannot be solved this way. For instance, the efficiency of object recognition systems may be judged using the criterion of "stability and sensitivity" (Marr and Nishihara 1978, p. 272). Accordingly, descriptions must reflect the similarity of objects thus enabling generalization. At the same time subtle differences need to be preserved to allow discrimination. Stable information representing global aspects of object shape must be decoupled, therefore, from information representing finer details. This can be achieved by relying on prominent pattern components for similarity judgments, whereas full pattern representations are used for discrimination (Rentschler et al. 1996).

More generally, traditional pattern recognition works well for simple isolated patterns but is inadequate for complex patterns and objects embedded in scenes. Image interpretation by computer therefore relies on the extraction of features of image parts and features of part relations that are linked together to form structural descriptions. Sets of hierarchically organized rules ("graphs") are then generated for classification to the extent needed for solving a given recognition problem. Classification performance can be improved further by feeding back results from rule evaluation to earlier stages of the rule generation system. Such methodologies of syntactic pattern recognition (see Caelli and Bischof 1997) have been adapted to the analysis of human image understanding (Rentschler and Jüttner 2007; see also the chapter by M. Jüttner, this volume) and object recognition (Osman et al. 2000). That approach would seem to be particularly appropriate for implementing cognitive functions as it integrates bottom-up and top-down processing characteristics. However, the various degrees of freedom of implementing such systems warrant further experimental research into the roles of prior knowledge and learning in human 3D object recognition.

We therefore sought to distinguish representations and recognition using a psychophysical paradigm of category learning involving priming. Priming is a technique from memory research using the beneficial influence of pre-exposure to a stimulus in the absence of explicit instructions to remember the stimulus (Biederman and Cooper 1991; Cooper et al. 1992). When used in combination with an invariant procedure of recognition involving fixed stimulus sets, any effect of priming must be attributed to object memory, i.e., representation.

3 Learning 3D Structure from Images

Our recognition paradigm used two sets of 3D objects consisting of one bilaterally symmetric object and one pair of handed (left and right) objects each (Fig. 1). Following priming (Fig. 2), participants were trained to classify a set of 22 learning views (Fig. 3). Upon reaching 90% correct, participants classified 83 test views (64 *novel* views, 19 *learned* views). Classification performance was measured in terms of signal detection accuracy (d prime; see Rentschler et al. 2004) and response time.

In the first experiment (Gschwind et al. 2004), we used objects built from spheres termed *spheres.* Resulting views were poor in ordered feature elements

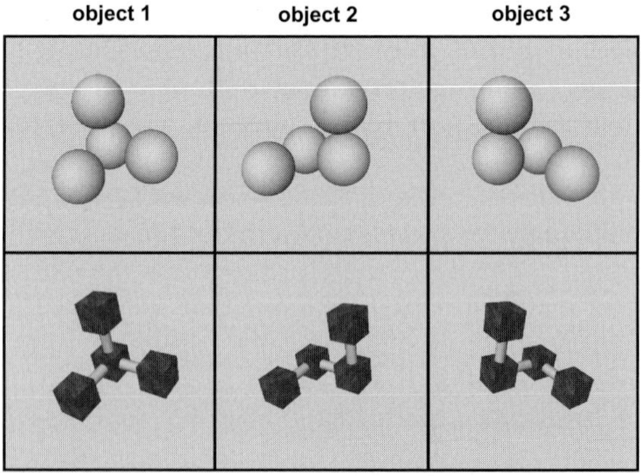

Fig. 1. Test sets of one bilaterally symmetric 3D object and one pair of handed (mirror symmetric) 3D objects. Each object was composed of four spheres (top) or cubes connected by rods (bottom). Three object parts formed an isosceles rectangular triangle, while the fourth one was placed perpendicularly above the centre of one of the base parts. Objects were generated both as physical models and virtual models. Physical models were constructed of polystyrene balls (6 cm diameter) or wooden cubes (3 cm sides) and rods (3 cm long, 1.2 cm diameter). Virtual models were generated and displayed as perspective 2D projections by the Open Inventor^{TM} (Silicon Graphics, Inc.) 3D developer's toolkit. A lighting model of mixed directed and diffuse illumination and a lack of cast shadows was used

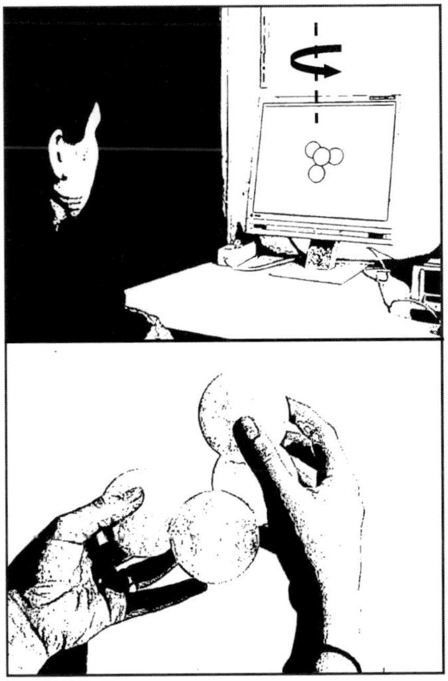

FIG. 2. For *vision* priming (*top*), participants watched one after the other computergraphic projections of the 3D objects successively rotating around the three principal axes. Two cycles of exposure of 90 s and 10 s per axis were used. For *motor* priming (*bottom*), the blindfolded subjects manipulated the physical models without restriction. No instructions other than the invitation to familiarize themselves with the objects were given. Priming lasted for 5 min and was followed by category learning

and connectivity of parts. This raised a question regarding the extent to which priming effects depended on stimulus information. We sought to answer this question in the second experiment using a set of modified stimuli termed *cubes*. The latter set had the same macrogeometric structure as *spheres* but textured cubes and rods as parts (see Fig. 1). The conditions of generating learning and test views, priming, as well as category learning and generalization were identical for both experiments.

Figure 4 shows the effects of priming in terms of classification performance in the first unit of category learning. With *spheres*, priming did not significantly affect the accuracy for object 1, perhaps because subjects were already at ceiling. Yet *motor* priming significantly improved classification of the handed objects 2 and 3 (Fig. 4, top left). For *cubes* (Fig. 4, top right), both *motor* and *vision* priming were equally effective in inducing classification, with the induction effect being most pronounced for non-handed object 1. Response times tended to be increased by *vision* and *motor* priming for the classification of *spheres* (Fig. 4, bottom left), although significance was only reached with *motor* priming for non-handed object

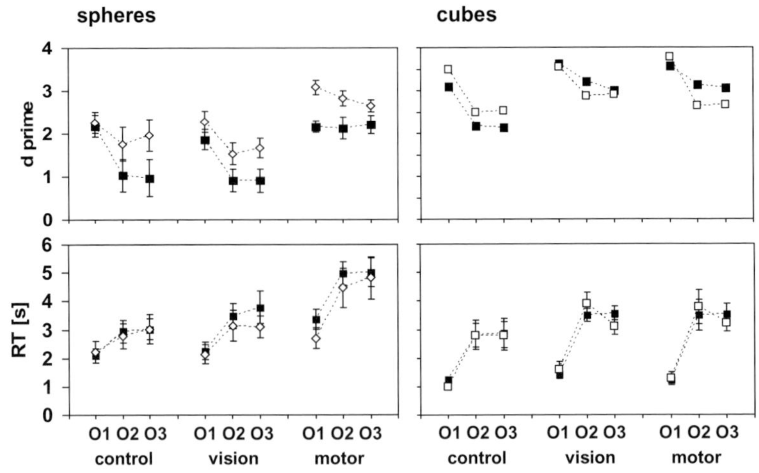

FIG. 5. Generalization to novel viewpoints for *spheres* (left) and *cubes* (right). Accuracies (d prime) cumulated over views, classification decisions, and observers at the top, corresponding response times at the bottom. Closed symbols denote generalization performance from novel viewpoints, open symbols from learned viewpoints. Each view was presented three times. Error bars: ± 1 S.E. (novel viewpoints: N = 7∗3∗16 control/object1, N = 7∗3∗24 control/objects23, N = 10∗3∗16 vision/motor/object1, N = 10∗3∗24 vision/motor/objects23; learned viewpoints: N = 7∗3∗5 control/object1, N = 7∗3∗7 control/objects23, N = 10∗3∗5 vision/motor/object1, N = 10∗3∗7 vision/motor/objects23). No error bars are given for the classification accuracies for *cubes* stimuli because of deviation from normal distribution

significant effect of priming conditions nor a significant difference between learned and novel views. That is, with *cubes* there occurred no differences in recognition performance between object types, priming conditions, or learned and novel views.

5 Inverse Problems and Spatial Transformations

Non-discrimination of handed objects is predicted by structural recognition models using non-directed part relations (e.g., Hummel 2001) and by view-based recognition models (e.g., Riesenhuber and Poggio 1999). Moreover, with *spheres* we found performance to be view-dependent consistent with the predictions of view-based recognition models. This would seem to support the "rotation-for-handedness" hypothesis (Tarr and Pinker 1989; Tarr 1995; Gauthier et al. 2002), according to which humans rely for recognition on reflection-invariant mechanisms in the brain and use mental rotation (Shepard and Metzler 1971) to disambiguate handedness.

The latter hypothesis, however, meets two difficulties when applied to the results of our experiments. First, images are generally ambiguous with regard to the 3D structures they are derived from. The solution for such inverse problems critically depends upon the operation of constraints, i.e., prior object knowledge (Pizlo 2001). This is why in previous studies on mental rotation subjects were given "a great deal of feedback about the 3D structure of each object" (Tarr 1995, p. 61). Our observers had no prior object knowledge under the *control* condition and were found to be completely unable, both for *spheres* and *cubes*, to disambiguate handedness early in practice (Fig. 4, top). Second, we used 2D views generated by conjointly varying the three Euler angles of rotation (see Korn and Korn 1968, Section 14.10). To reliably recover these angles from a given 2D view is impossible, and the rotation of the image plane was an additional source of uncertainty for the matching process. On these grounds, we reject the rotation-for-handedness hypothesis according to which our observers could have achieved disambiguation of handedness by employing continuous rotations around specific axes in 3D. Instead, for both non-handed and handed objects, they must have iteratively used combinations of spatial transformations.

Consistent with the latter conclusion, subjects with *motor* priming needed for the recognition of both non-handed and handed *spheres* prolonged response times, typically moved their hands during classification, and spontaneously reported having mentally rotated the candidate models for classification. The retardation of response times would seem to reflect, therefore, the times needed for generating internalized candidate models, transforming them during the matching process, and executing additional transformations to align mirror symmetric counterparts. This implies that, for *spheres* with *motor* priming, our recognition paradigm could not be separated into one of recognizing the non-handed object and one of discriminating handedness. Indeed, the improvement of category learning through motor priming was most pronounced for non-handed object 1. The signal detection analysis of data from the generalization phase demonstrated that this resulted at least partly from a reduction of the misclassification of views of the handed objects as views of the non-handed object.

We then turn to the question of how *motor* priming facilitated the classification of stimuli built from spheres. Clearly, such type of priming drew the attention of subjects to the third stimulus dimension. This enabled them to explicitly generate relational 3D representations (Thoma et al. 2004). Participants may have solved the inverse problem for *spheres* by encoding temporal sequences of exploratory finger and hand movements along the *physical* object models. As object palpation directly evokes mental imagery (Critchley 1953, chap. IV), it seems that some sort of kinetic object traces were stored in multimodal representations (e.g., Zangaladze et al. 1999). Subjects may then have inferred the connectivity of sphere parts, i.e., 3D structure, from linking object parts exposed in 2D views to such internalized representations. Conversely, we suggest that the type of prior knowledge provided by *vision* priming did not allow the solution of the inverse problem for *spheres*. Indeed, during *vision* priming subjects noted ambiguous rotation-in-depth of the *spheres* objects. These effects were caused by uncertainties of correspondence between object views displayed during motion.

6 Role of Image Understanding in Invariant Recognition

From the equivalence of *vision* and *motor* priming for classifying *cubes* (Fig. 5), we conclude that the clear connectivity between parts and the related ordering of feature elements helped the solution of the inverse problem right from the visual stimulus. Moreover, the parallel contours of cube parts facilitated matching thus supporting the verification of candidate 3D object models. Therefore, the classification of *cubes* would seem to be an instance of fast and accurate recognition that is viewpoint invariant as predicted by RBC. Indeed, for *cubes* we found recognition performance to be view-invariant. Furthermore, the classification of handed objects built from *cubes* entailed prolonged response times, thus indicating the need of aligning internalized object models to an external reference system.

In case of objects built from *spheres*, the extraction of part relations from 2D views was difficult. The parts as such left the axes of connectivity between them completely unspecified. The image understanding of the observers therefore benefited greatly from structural cues obtained from motor memory, thus presumably using 2D representations augmented by 3D information from motor memory (see Liu et al. 1995). The matching of such reduced object models to input data, however, entailed an increase in search complexity, i.e., the amount of spatial transformations and matching needed for categorization. As a result, the response times for classifying both types of objects built from spheres, non-handed and handed, were prolonged.

These findings emphasize the role of image understanding in object recognition. The two sets of objects had identical structural characteristics relevant for classification, and their respective members were readily decomposed into identical parts. Object recognition relied, therefore, entirely on the ability to recover part relations from 2D views.

7 Conclusions

We have shown that early in practice, humans were virtually blind to structural differences of 3D objects composed of identical sphere-shaped parts. Category learning improved recognition but more for non-handed objects than for handed objects. Prior knowledge from passively inspecting 2D views of depth-rotating objects did not affect recognition, whereas active haptic exploration of physical 3D models enabled equally accurate but view-dependent recognition of both non-handed and handed objects. Using objects with the same macrogeometrical features but clear connectivity of cube-shaped parts yielded very different results. Recognition was fast and accurate early in practice for the non-handed object.

Yet, with both types of prior knowledge, category learning enabled equally accurate and view-independent recognition for both non-handed and handed objects.

These results demonstrate, on the one hand, that there is no absolute difference between stimuli that allow distinct structural descriptions for 3D object recognition and stimuli that do not (e.g., Biederman and Gerhardstein 1993). Prior knowledge and learning play an important role in determining the extent to which image regions and their relations can be referenced to mental object representations. On the other hand, the structure-based recognition of 3D objects is not accommodated by the multiple-views theory of recognition (e.g., Bülthoff and Edelman 1992). These observations would seem to be consistent with the conclusions by Christou and Bülthoff (2000), according to whom the nature of object representations depends on whether there is enough stimulus information for the recognition task at hand.

We therefore propose that observers build 3D representations for object recognition as long as sufficient stimulus information and prior knowledge are available. Yet internalized 3D models may be too similar to allow their disambiguation concerning class membership, a situation typically encountered in classification at the subordinate level. Alternatively, observers may fail early in practice to extract from input images view-invariant geometric primitives in distinct relations. Category learning might then enable them to derive such structural descriptions. Otherwise, they would resort to the use of object representations in image format and corresponding matching behavior, thus increasing classification performance for learned views at the expense of decreased performance in generalization to novel views.

Acknowledgment. This chapter benefited greatly from the reviews and comments of Irving Biederman, Terry Caelli, Martin Jüttner, and Zili Liu.

References

Ballard DH, Brown CM (1982) Computer vision. Prentice Hall, Englewood Cliffs NJ

Biederman I (1987) Recognition-by-components: a theory of human image understanding. Psychol Rev 94:115–147

Biederman I, Cooper EE (1991) Priming contour-deleted images: evidence for intermediate representations in visual object recognition. Cognit Psychol 23:393–419

Biederman I, Gerhardstein PC (1993) Recognizing depth-rotated objects: evidence and conditions for three-dimensional viewpoint invariance. J Exp Psychol 19:1162–1182

Bülthoff H, Edelman S (1992) Psychophysical support for a two-dimensional view interpolation theory of object recognition. Proc Natl Acad Sci USA 89:60–64

Caelli T, Bischof WF (1997) Machine learning and image interpretation. Plenum Press, New York

Christou C, Bülthoff HH (2000) Perception, representation and recognition: a holistic view of recognition. Spat Vis 13:265–276

Cooper LA, Schacter DL, Ballesteros S, Moore C (1992) Priming and recognition of transformed three-dimensional objects: effects of size and reflection. J Exp Psychol Learn Mem Cogn 18:43–57

Critchley M (1953) The parietal lobes. Edward Arnold, London

Dickinson SJ (1993) Part-based modeling and qualitative recognition. In: Jain AK, Flynn PJ (Eds) Three-dimensional object recognition systems. Elsevier, Amsterdam, pp 201–228

Edelman S, Bülthoff HH (1992) Orientation dependence in the recognition of familiar and novel views of 3D objects. Vision Res 32:2385–4000

Gauthier I, Hayward WG, Tarr MJ, Anderson AW, Skudlarski P, Gore JC (2002) BOLD activity during mental rotation and viewpoint-dependent object recognition. Neuron 34:161–171

Gschwind M, Brettel H, Osman E, Rentschler I (2004) Structured but view-dependent representation for visual 3-D object classification. Perception 33(Suppl):73

Haykin S (1999) Neural networks. Prentice Hall, Upper Saddle River NJ

Hummel JE (2001) Complementary solutions to the binding problem in vision: implications for shape perception and object recognition. Vis Cogn 8:489–517

Hummel JE, Biederman I (1992) Dynamic binding in a neural network for shape recognition. Psychol Rev 99:480–517

Johnson KE, Mervis CB (1997) Effects of varying levels of expertise on the basic level of categorization. J Exp Psychol Gen 126:248–277

Korn GA, Korn TM (1968) Mathematical handbook for scientists and engineers. McGraw-Hill, New York, Section 14.10

Liu Z (1996) Viewpoint dependency in object representation and recognition. Spat Vis 9:491–521

Liu Z, Knill DC, Kersten D (1995) Object classification for human and ideal observers. Vision Res 35:549–568

Liu Z, Kersten D, Knill DC (1999) Dissociating stimulus information from internal representation – a case study in object recognition. Vision Res 39:603–612

Marr D, Nishihara HK (1978) Representation and recognition of the spatial organisation of three-dimensional shapes. Proc R Soc Lond B 200:269–294

Osman E, Pearce AR, Jüttner M, Rentschler I (2000) Reconstructing mental object representations: a machine vision approach to human visual recognition. Spat Vis 13:277–286

Pizlo Z (2001) Perception viewed as an inverse problem. Vision Res 41:3145–3161

Rentschler I, Jüttner M (2007) Mirror-image relations in category learning. Vis Cogn 15:211–237

Rentschler I, Barth E, Caelli T, Zetzsche C, Jüttner M (1996) Generalization of form in visual pattern classification. Spat Vis 10:59–85

Rentschler I, Jüttner M, Osman E, Müller A, Caelli T (2004) Development of configural 3D object recognition. Behav Brain Res 149:107–111

Riesenhuber M, Poggio T (1999) Hierarchical models of object recognition in cortex. Nat Neurosci 2:1019–1025

Shepard RN, Metzler J (1971) Mental rotation of three-dimensional objects. Science 171:701–703

Tanaka JW, Taylor M (1991) Object categories and expertise: is the basic level in the eye of the beholder? Cognit Psychol 23:457–482

Tarr M (1995) Rotating objects to recognize them: a case study of the role of viewpoint dependency in the recognition of three-dimensional objects. Psychonom Bull Rev 2:55–82

Tarr MJ, Pinker SM (1989) Mental rotation and orientation dependence in shape recognition. Cognit Psychol 21:233–282

Thoma V, Hummel JE, Davidoff J (2004) Evidence for holistic representations of ignored images and analytic representations of attended images. J Exp Psychol 30:257–267

Tjan BS, Legge GE (1998) The viewpoint complexity of an object-recognition task. Vision Res 38:2335–2350

Zangaladze A, Epstein CM, Grafton S, Sathian K (1999) Involvement of visual cortex in tactile discrimination of orientation. Nature 401:587–590

8
Neural Representation of Faces in Human Visual Cortex: the Roles of Attention, Emotion, and Viewpoint

Patrik Vuilleumier

1 Introduction

Faces constitute a special class of visual stimuli not only because we possess expert visual skills and specialized brain areas to recognize them, but also because we can extract a rich set of socially and affectively important information from them in a seemingly effortless manner. Abundant research conducted in cognitive psychology, neuroscience, and clinical neuropsychology has provided an elaborate model of the complex functional architecture underlying these different aspects of face processing, each presumably associated with specific neural substrates that are interconnected all together within a large-scale distributed network (Grüsser and Landis 1991). Thus, many influential neurocognitive models have proposed that face recognition may proceed along a series of distinct stages organized in a hierarchical stream of processing (Bruce and Young 1986; Haxby et al. 2000), from low-level visual analysis subserving the detection and organization of facial features, up to higher-level processes allowing the storage and retrieval of personal information and other associative functions (Fig. 1a). Furthermore, some dissociations in recognition performance in healthy subjects, as well as neuropsychological deficits observed in patients with focal brain lesions, have led to the idea that different processing pathways might be responsible for extracting identity-related information versus other facial features related to emotional expression, eye gaze direction, or speech lip motion, and that such pathways might operate in parallel (Bruce and Young 1986; Grüsser and Landis 1991). To what extent these different processing streams may interact to influence each other, and how the different kinds of information may eventually be unified in a single face percept, are two fundamental questions that still remain to be determined.

Laboratory for Neurology and Imaging of Cognition, Department of Neurosciences, University Medical Center (CMU), 1 rue Michel-Servet, 1211 Geneva; Department of Clinical Neurology, Unviersity Hospital (HUG), 24 rue Micheli-Du-Crest, 1211 Geneva; Swiss center for affective sciences, Department of Psychology, University of Geneva, 7 rue des Battoirs, 1205 Geneva, Switzerland

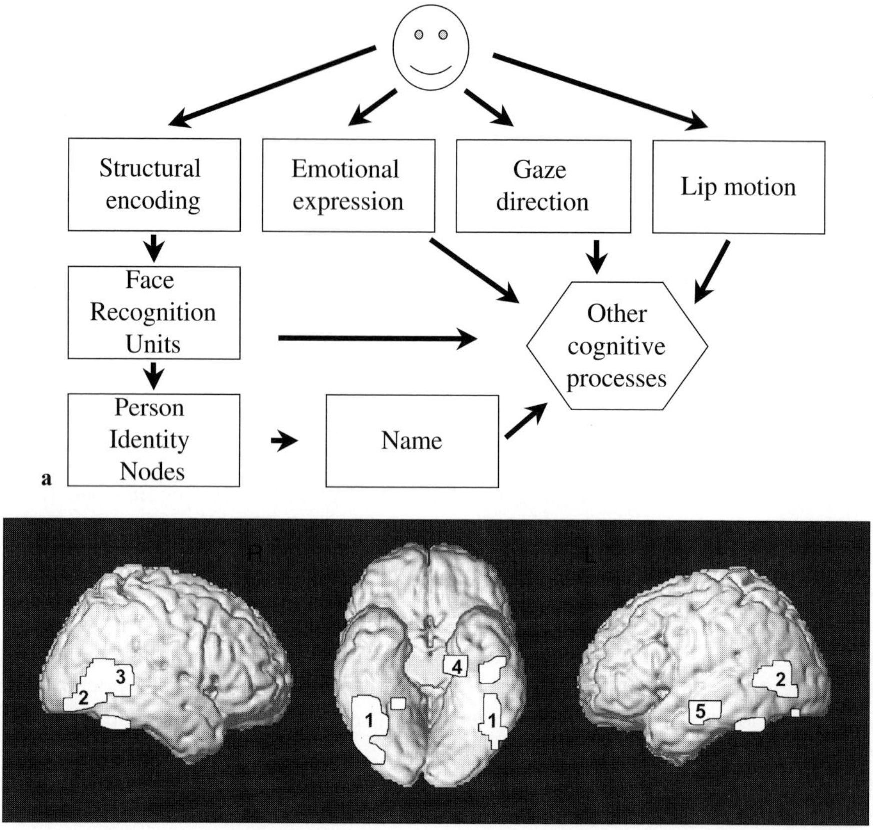

FIG. 1. **a** Traditional cognitive model of face processing derived from Bruce and Young (1986), in which identity and expression are processed along pathways of separate serial stages. **b** Network of brain areas typically activated by contrasting faces > other visual objects in fMRI, bilaterally but with variable hemispheric asymmetry, including (1) fusiform cortex, FFA; (2) lateral occipital cortex, OFA; (3) superior temporal sulcus, STS; (4) amygdala; (5) anterior lateral temporal cortex

Recent progress in functional brain imaging has allowed a tremendous refinement of our knowledge of the anatomy of the human face recognition system, and its operating properties. In particular, a cortical region in the human fusiform gyrus has been identified as critically implicated in face processing across a variety of studies using positron emission tomography (PET) (Sergent et al. 1992; Haxby et al. 1994) or functional resonance imaging (fMRI) (Kanwisher et al. 1997; McCarthy et al. 1997). This region is commonly referred to as the "fusiform face area" (FFA), and generally thought to play a major role in the detection as well as the discrimination of individual faces (Gauthier et al. 2000; Grill-Spector et al. 2004). The FFA is consistently activated by pictures or line-drawings of human faces more than by any other categories of visual objects or scenes, and its activation correlates with face perception during the presentation of ambiguous stimuli, perceptual rivalry, or even mental imagery. However, several other brain regions, within and outside the visual system, are also differentially activated by faces relative to other visual objects (Sergent et al. 1992; Haxby et al. 2000). These regions include the lateral occipital face area (OFA), the superior temporal sulcus (STS), the amygdala, plus other areas in the temporal poles and ventromedial prefrontal cortex (Fig. 1b). In accord with previous cognitive models (Bruce and Young 1986), it has been proposed that the FFA might be crucially involved in processing visual features carrying face identity information, which should remain relatively invariant across changes due to expression, viewpoint, or pictorial format. Conversely, STS and amygdala might be more important for processing changing or dynamic features in faces, such as expression or gaze, which are socially and emotionally relevant and shared across many different identities (Haxby et al. 2000). The role of other brain regions still remains largely unsettled (for extended neuroanatomical model, see Gobbini and Haxby 2007).

However, although there is now abundant evidence that face identity is processed in the FFA and that facial expression is processed in amygdala and STS, there is also increasing evidence that these two aspects of face recognition might not be entirely encapsulated and separately implemented in these different regions, as previously proposed by cognitive models. In particular, the present chapter will focus on two series of recent brain imaging studies showing that face representation in the FFA is not totally insensitive to emotional expression and not totally independent from viewpoint. By illustrating how different regions in the face recognition system may not carry out specialized processes alone but dynamically interact with each other, these findings call for a refinement of the current neurocognitive models of face recognition, which have considered only a serial feedforward mode of information processing but ignored the role of more interactive and re-entrant mechanisms.

2 Emotional Influences on Face Processing in Fusiform Cortex

A number of brain imaging studies have consistently shown that the activation of sensory cortical areas can be enhanced for emotionally relevant stimuli, including not only faces (Morris et al. 1998a; Vuilleumier et al. 2001) but also pictures (Lane et al. 1999; Sabatinelli et al. 2005) or voices (Grandjean et al. 2005). For instance, such increases may arise in the visual cortex with faces displaying fearful relative to neutral expressions, or with photographs containing aversive relative to more mundane scenes. A negative emotional content generally appears much more efficient in producing such increases, particularly for faces (Surguladze et al. 2003), although positive arousal can sometimes produce similar effects (Mourao-Miranda et al. 2003; Sabatinelli et al. 2005).

Such increases in response to emotional (e.g., fearful) faces have been observed in various regions such as the fusiform cortex, posterior inferior and lateral temporal cortex, as well as in very early occipital areas such as the primary striate cortex (area V1) (Morris et al. 1998a; Vuilleumier et al. 2001; Pessoa et al. 2002b). However, these effects also exhibit a relative selectivity depending on the category of the emotional stimulus. For instance, in an fMRI study (Vuilleumier et al. 2001), where pictures of faces with either a fearful or neutral expression were presented together with pictures of houses, fear-related increases were found to arise selectively in the lateral fusiform region that also showed face-specific responses, corresponding to the FFA. However, a nearby region in the parahippocampal cortex showing house-specific responses (i.e., the parahippocampal place area, PPA) was not modulated by the emotional expression of faces seen with the houses (Fig. 2). This finding suggests that emotional signals received from faces can produce a selective influence on the cortical representation of faces in the FFA, and that face identity processing in fusiform cortex may not be purely encapsulated and immune to interactions with processes involved in face expression recognition.

Moreover, the modulation of the FFA by emotional expression of faces was found to arise in the same voxels in the cortex as the modulation produced by selective attention to faces (Fig. 2). In the same fMRI study using faces and houses presented together (Vuilleumier et al. 2001), we could compare the effect of expression and the effect of selective attention by manipulating attention and emotion orthogonally, while keeping the task identical across all conditions. While visual arrays always contained two faces (fearful or neutral expression) and two houses, the observers had to concentrate on two pictures only (either the vertical or horizontal pair) on each single trial, in order to make same/different judgments for these two pictures. Thus, we could measure the differential impact on neural responses due to fearful *vs* neutral emotional expressions when faces were either in the focus of attention, or outside the focus of attention. Three major results were found. First, the effects of emotion and attention on FFA responses were additive to each other, with a similar enhancement to fearful expression when faces were in the focus of attention (for a same/different judgment) and when they were outside the focus of attention (with a same/different judgment being made on houses instead). Second, the effect of emotion from ignored faces arose in the FFA despite a strong reduction in activity due to inattention when observers concentrated on the houses. Third, the peak of emotional effects in the FFA was exactly the same as the peak of attentional effects, and fully consistent with the location of face-selective areas reported in previous studies. This pattern of results has then been replicated in two further fMRI studies using the same paradigm in different subjects (Bentley et al. 2003; Vuilleumier et al. 2004).

Taken together, these findings suggest that FFA activity may be controlled by top-down influences imposed not only by attentional systems (presumably mediated by fronto-parietal cortical networks), based on current task demands (Wojciulik et al. 1998), but also by emotional systems extracting the potential affective or social value of faces even when these are not currently task-relevant or in the focus of attention. Such emotional effects on neural responses of the FFA might result in a more salient representation of faces with particular affec-

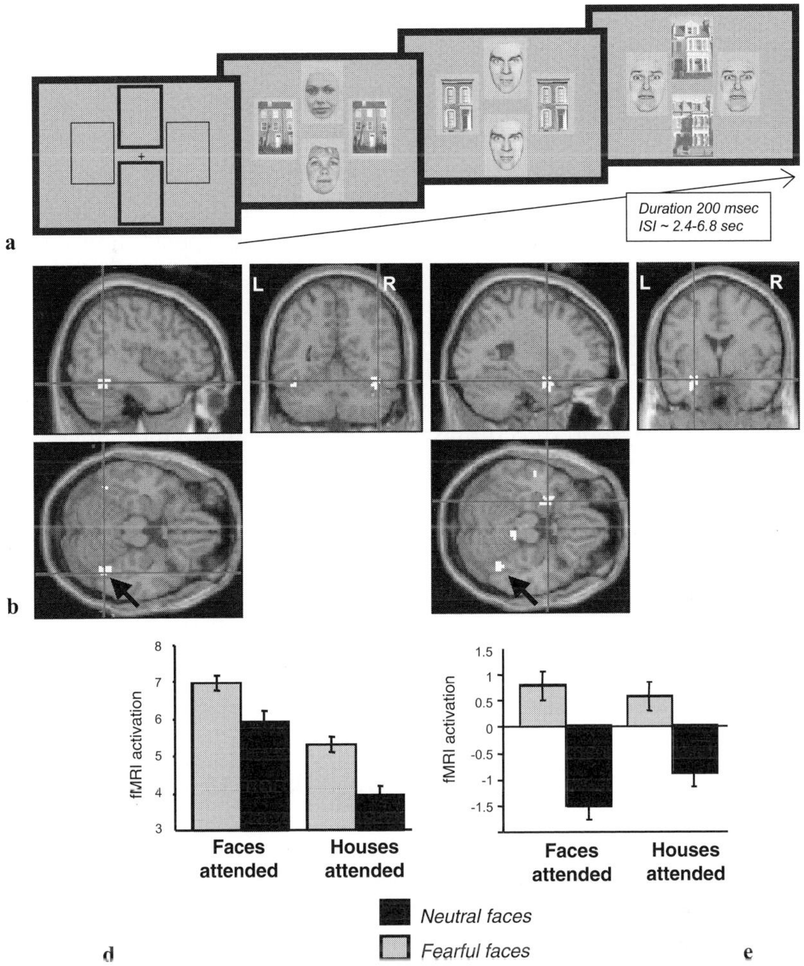

FIG. 2. **a** Paradigm used to compare the effects of emotion and attention in responses to faces. On each trial, two faces and two houses are presented together, aligned in either vertical or horizontal pair, while observers are instructed by an initial cue to concentrate only on one pair of locations (here vertical). Faces can be neutral or fearful. **b** Effect of attention to faces versus houses, resulting in an increased activation of both right and left FFA. **c** Effect of fearful versus neutral expression in faces, resulting in a similar increase in the FFA (bilaterally but stronger on the right, see arrow), in addition to an activation of the amygdala (bilaterally but stronger on the left as shown here). Average parameter estimates of activity (±SE) are shown across all conditions of attention and expression for (**d**) right FFA and (**e**) left amygdala

tive values, such as threat, and thus provide a plausible neural substrate for attentional biases towards emotional faces, as observed in several behavioral studies (Vuilleumier and Schwartz 2001a, b; Fox 2002; Vuilleumier 2005). For instance, as compared to neutral faces, faces with threat-related expressions tend to produce faster detection in visual search (Fox et al. 2000; Eastwood et al. 2001) or visual orienting paradigms (Mogg et al. 1994; Pourtois et al. 2004).

Our fMRI data also accord with neurophysiological recordings in the monkey showing that some face-selective neurons in temporal cortex may show enhanced responses to faces with particular expressions (Sugase et al. 1999). However, in neurophysiological recordings, other face-selective neurons in the same cortical area may also show enhanced responses to faces with a particular identity. Unfortunately, the spatial resolution of fMRI is still insufficient to determine whether distinct neuronal populations in the human FFA might be sensitive to facial expression or identity, and thus differentially modulated by emotion and attention. In the future, higher-field MRI and voxel-by-voxel analysis of activated regions within fusiform cortex might provide better insights into the fine cortical organization of distinct neuron clusters with different processing preferences. Some recordings in STS in the monkey have shown that identity-selective and emotion-selective neurons might be arranged in distinct clusters along the ventral and dorsal banks of STS, respectively (Hasselmo et al. 1989). However, it is still unclear what is the homology between these cortical visual areas in monkeys and humans.

3 Distant Sources of Emotional Signals from the Amygdala

Interestingly, neurophysiological data in the monkey suggest that an emotional modulation of face processing in visual cortex might occur only after some delay following the initial face-selective responses. Thus, the first neuronal activity (<100 ms) might primarily code for global stimulus category (face *vs* other object) whereas subsequent activity (100–150 ms) might code for finer information such as expression and/or identity (Sugase et al. 1999). This delayed modulation has therefore been attributed to some re-entrant influences from distant brain areas responsible for processing affective or familiarity information. In particular, emotional influences on visual cortex might be provided by the amygdala, which is known to be critically implicated in emotional processing, especially threat, and to give rise to feedback projections to all levels of the ventral visual cortical stream (Amaral et al. 2003). These anatomical connections might allow the amygdala to have substantial modulatory control over sensory processing at several stages along the visual pathways.

In agreement with this idea, our fMRI results revealed that the amygdala could respond to fearful faces irrespective of whether observers had to concentrate on faces or houses (Vuilleumier et al. 2001). Thus, amygdala activation was not significantly influenced by attention in this paradigm, despite the robust effect of attention on visual cortex (Fig. 2). These data suggest that emotional responses in the amygdala may not rely on face processing taking place in the fusiform cortex, consistent with other findings that the amygdala can still be activated by threat cues in some conditions when observers are not aware of these cues (e.g., during masking (Morris et al. 1998b; Whalen et al. 1998), rivalry (Pasley et al. 2004; Williams et al. 2004), or blindsight (Morris et al. 2001; Pegna et al. 2005)). Yet it is possible that the amygdala responses can also be influenced by attention in other conditions (Pessoa et al. 2002a, b). More importantly, these results also suggest that amygdala activation to fearful expression might provide the primary source of emotional modulation on the FFA, leading to the persistent and additive enhancement regardless of the concomitant attentional modulation.

To test directly this idea of amygdala influences on the FFA, we conducted another fMRI study using the same paradigm with face-and-house pairs as above, but now in patients with amygdala lesions (Vuilleumier et al. 2004). In this study, two groups of patients with medial temporal lobe sclerosis were compared, half in whom the lesions affected both the amygdala and hippocampus, and the other half in whom the lesions affected the hippocampus only and spared the amygdala. Patients with hippocampus damage but intact amygdala showed a normal increased activation for fearful faces in fusiform and occipital cortex, whereas patients with additional amygdala damage showed no differential responses to fear in the FFA. In addition, parametric analyses revealed a linear inverse correlation between the severity of amygdala sclerosis and the enhancement of ipsilateral fusiform activity by fear, consistent with amygdala connections projecting mostly to ventral visual cortical pathways within the same hemisphere (Amaral et al. 2003). By contrast with this lack of emotional effects, both groups of patients showed a normal modulation of the FFA by attention to faces as compared to attention to houses. These findings therefore strongly support the idea that the amygdala can influence activity in distant visual areas and boost the representation of faces in the FFA based on their affective significance.

Face processing in the FFA is therefore likely to be partly controlled by "feedback" or re-entrant signals from the amygdala (Vuilleumier 2005), in addition to concomitant influences from other control systems in fronto-parietal attentional networks and probably still other sources yet to be identified. These modulatory influences from the amygdala may facilitate the detection of affectively significant information and enhance attention towards these salient stimuli, but also modify the establishment or retrieval of memory traces associated with emotional faces. In agreement with a role in detection and attention, previous behavioural results have shown that amygdala lesions in humans will abolish the typical attentional biases towards stimuli with threat versus neutral meaning. However, the functional consequences on memory still remain to be fully explored.

4 Distinct Visual Cues for Processing Faces in Fusiform Cortex and Amygdala

The fact that the amygdala might still respond to fearful faces presented outside the focus of attention (Vuilleumier et al. 2001), or sometimes even outside awareness (Morris et al. 1998b; Whalen et al. 1998; Pasley et al. 2004; Williams et al.

2004), has commonly been explained by the existence of distinct neural pathways for processing emotional cues. In particular, based on animal studies of fear-conditioning (LeDoux 2000) and studies of patients with blindsight after destruction of their primary visual cortex (Morris et al. 2001; Pegna et al. 2005), it has been hypothesized that the detection of threat-related stimuli might not depend on elaborate cortical analysis but rather implicate a fast subcortical pathway conveying only "quick and dirty" signals (Morris et al. 1999; LeDoux 2000). This subcortical pathway might involve direct visual inputs to the superior colliculus and/or pulvinar nucleus of the thalamus, bypassing early cortical stages of processing from geniculo-striate pathways to the ventral occipito-temporal stream (Morris et al. 1999, 2001). However, although this subcortical route might play an important role in blindsight or cortical blindness, its connections to the amygdala still remain controversial in humans (Pessoa 2005), and "quick and dirty" information might also reach the amygdala through a first volley of bottom-up inputs within the visual cortex prior to full perceptual analysis and attentional selection (Vuilleumier 2005).

In any case, a preservation of amygdala activation to stimuli perceived under poor conditions of visibility would make sense in order to afford rapid and efficient response to threat. Moreover, subcortical visual pathways are known to carry only crude visual information with low-spatial frequency, extracted from magnocellular pathways, whereas finer visual information in high-spatial frequency from parvocellular pathways project exclusively to cortical areas in the ventral occipito-temporal stream (Merigan and Maunsell 1993; Sahraie et al. 2002). Using fMRI in healthy subjects, we therefore tested for any differential sensitivity of amygdala and fusiform cortex to low-spatial frequency (LSF) and high-spatial frequency (HSF) (Vuilleumier et al. 2003a). Observers were presented with photographs of faces displaying either a neutral and fearful expression, and containing either low-pass, high-pass, or intact (broad-band) spatial frequency content (Fig. 3). Activation of the FFA was found to be generally reduced for LSF faces relative to intact or high-pass faces, irrespective of expression, consistent with an important role of fine edge and texture information in driving activity of temporal visual cortex. By contrast, amygdala responses to fearful expression were greater for both LSF and intact faces than for HSF faces, despite the reduced response to HSF in the FFA.

This dissociation suggests that amygdala and FFA may extract different spatial-frequency content in faces, which may play distinct roles in expression and identity processing, respectively (Vuilleumier et al. 2003a). This would be consistent with behavioral studies showing different perceptual biases to LSF and HSF cues when observers must categorize the identity and expression of "hybrid" stimuli, in which different faces with different content are superimposed (Schyns and Oliva 1999).

Remarkably, however, we found that the FFA was increased by fearful relative to neutral expression only with LSF (and intact) faces, but not with HSF, even though the FFA was generally less sensitive to HSF than LSF cues (Vuilleumier et al. 2003a). This pattern provides further support to the idea that such

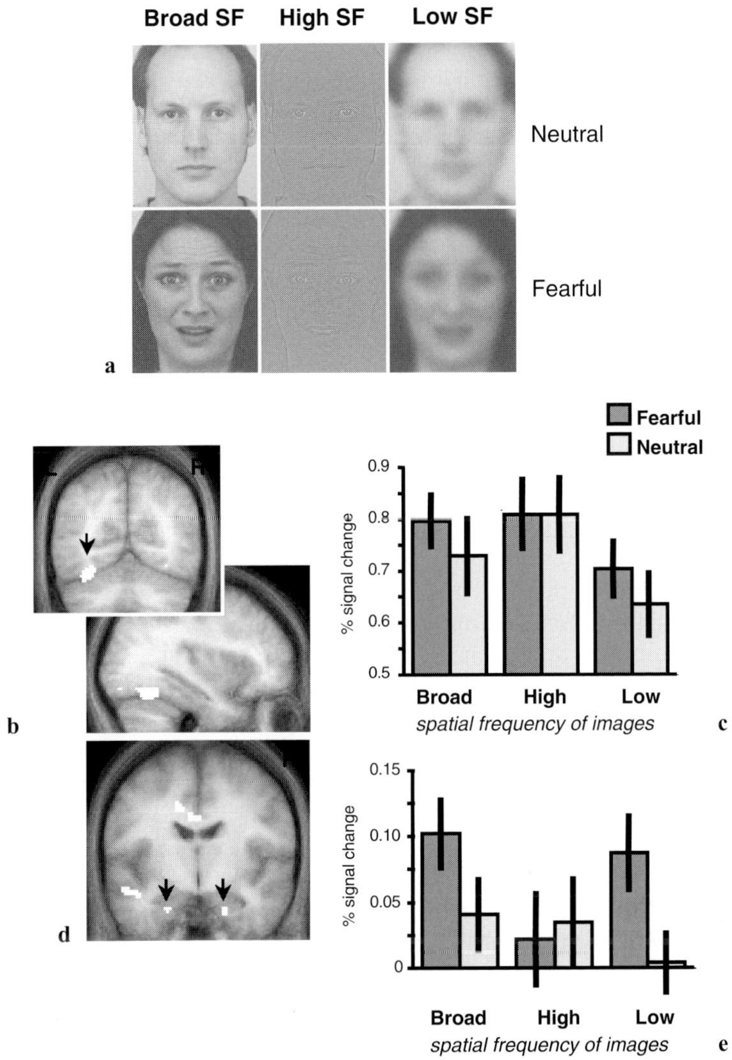

FIG. 3. **a** Stimuli used to compare face processing based on the low-spatial frequency (LSF) and high-spatial frequency (HSF) content of images, relative to normal (broadband) images. **b** Posterior fusiform cortex was activated by the presence of HSF in face stimuli, but not by LSF. **c** Average parameter estimates of activity (±SE) in FFA. Note however that an enhancement by fearful expression was driven by the presence of LSF. **d** Amygdala was activated by fearful expression in the LSF of face stimuli, but not by HSF. **e** Average parameter estimates of activity (±SE) in amygdala

emotional effect in FFA may depend on inputs from the amygdala, rather than on intrinsic cortical processing. The same pattern was observed in two subsequent imaging studies where "hybrid" faces were used. Both in fMRI (Winston et al. 2003) and ERPs (Pourtois et al. 2005a), we found that differential cortical responses to fearful *vs* neutral faces were evoked only when fearful expression was presented within the LSF content of pictures, irrespective of the expression of another superimposed face presented in HSF. This critical role of LSF information seems consistent not only with several recent studies showing that amygdala processing of fearful expression in faces may be highly sensitive to the large eye features that are typically present in these faces (Morris et al. 2002; Whalen et al. 2004; Adolphs et al. 2005), but also with some psychophysical results showing an important role of configural information for the recognition of face expression (rather than just local features) (Calder et al. 2000).

Conversely, our fMRI study (Vuilleumier et al. 2003a) also suggested that face identity processing in the FFA was established from HSF more reliably than from LSF cues. Because each individual face identity was repeated once during the whole course of the fMRI experiment, we could test for any repetition-priming effects induced by different visual images of the same face identity. Repetition-priming effects correspond to a selective decrease in the activation of cortical areas processing a particular stimulus type when this stimulus is repeated, relative to its first exposure, and such effects can thus reveal the specific attributes extracted by neurons in that particular area (Grill-Spector and Malach 2001; Naccache and Dehaene 2001). Here, by comparing repetition-priming effects for HSF and LSF faces relative to those for intact faces, we found that only faces first seen in HSF produced subsequent decrease when repeated later in a different format (Fig. 4), whereas faces first seen in LSF produced no decrease when repeated (Vuilleumier et al. 2003a). These data suggest that a long-term representation of identity in the FFA was more efficiently established and more efficiently generalized to other images when derived from HSF than from LSF information. Moreover, repetition-priming effects for identity across different images were found to predominate in more anterior regions of the fusiform cortex, whereas the peak of frequency-selectivity for HSF *vs* LSF was found in a more posterior fusiform region. Other imaging findings have also shown that the FFA might code for face identity irrespective of spatial frequency (Eger et al. 2004) or contrast polarity (George et al. 1999).

Taken together, these data suggest that face processing may not only take place in different brain pathways for different purposes (e.g., identity recognition in FFA and expression recognition in amygdala), but also exploit different information (e.g., LSF or HSF, global *vs* local cues) and probably proceed at different time-scale in different brain areas (with expression processed earlier in amygdala and then fed back to FFA). Thus, models of face processing should not only incorporate a "dual-route" framework for identity and emotion information (Bruce and Young 1986; Haxby et al. 2000), but also a "dual-stage" framework.

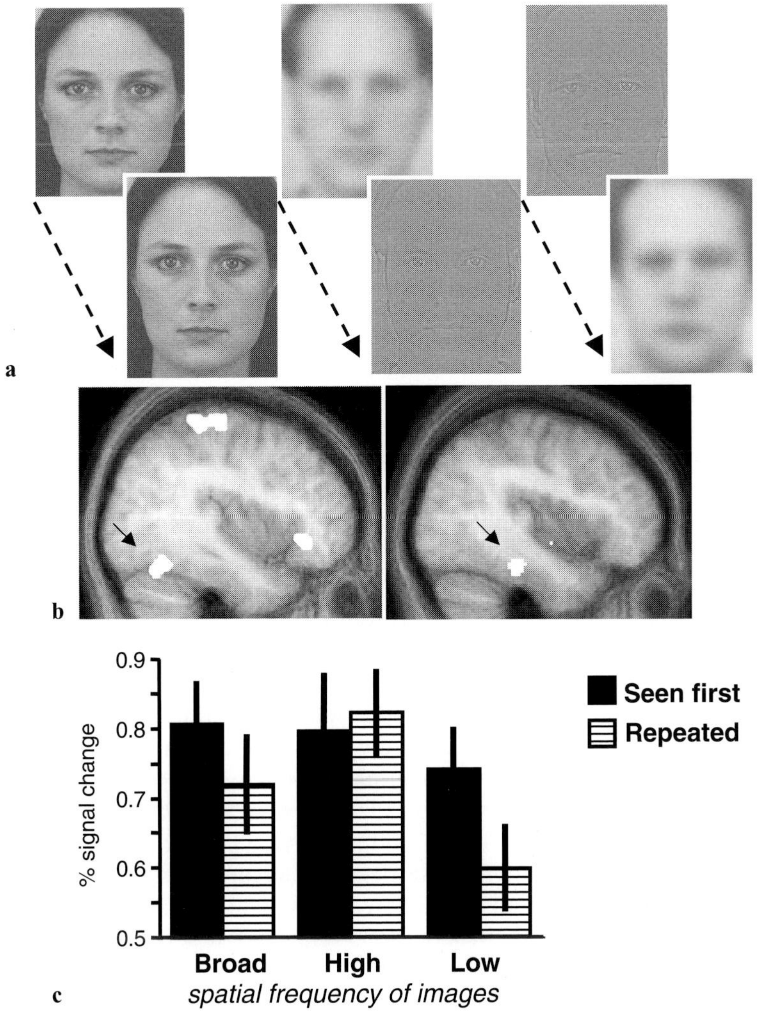

FIG. 4. **a** Stimuli used to test for repetition-priming effects when face identity is repeated, either in the same picture format or across different picture formats (e.g., first seen in LSF and later repeated in HSF, or vice versa). **b** Repetition-priming effects in posterior (left) and anterior (right) fusiform cortex, where responses showed a selective decrease when face identity was repeated irrespective of whether the repetition was with the same or with different images. **c** Average parameter estimates of activity (±SE) in anterior fusiform cortex, where repetition-priming were the strongest when the face identity was first seen in HSF and repeated in LSF (as opposed to the reverse order)

5 View-Selectivity and Invariance

If the FFA can encode faces irrespective of format and process identity across different spatial-frequency cues, what is the degree of invariance to other changes in visual inputs during identity recognition? A critical issue in visual perception in general has concerned how objects and faces can be identified despite changes in their visual appearance across different encounters (Biederman and Kalocsai 1997; Tarr and Bulthoff 1998; Biederman and Bar 2000; Vuilleumier et al. 2002). Thus, the identity of an individual face can usually be recognized across substantial visual changes due to different expressions, age, or viewpoint. In fact, we probably never see the same face twice with exactly the same view, yet we can readily identify a person across two meetings, or an old friend who has not been seen for several years. However, still little is known about how the visual system may achieve such efficient recognition abilities across very different visual inputs. Although the FFA has consistently been shown to process face identity cues (Gauthier et al. 2000; Grill-Spector et al. 2004), it remains unclear how face identity is represented in the FFA.

According to classic models of face recognition (Bruce and Young 1986), after some initial structural encoding stage, view-invariant traits might be extracted and stored into a long-term visual representation of a given individual face (e.g., "face recognition unit"), which may then allow a generalization of recognition from a particular view to another view of the same face. However, few studies have systematically examined whether the neural substrates of such "face recognition units" might correspond to the FFA and code for a particular face identity across different views (Grill-Spector et al. 1999). We have addressed this question in two recent brain imaging studies (Pourtois et al. 2005b, c) by using a repetition-priming paradigm in which different views of the same faces were presented twice, with an intervening delay of several minutes. As mentioned above, repetition-priming leads to a decreased activation for repeated stimuli as compared with their initial presentation, reflecting a selective adaptation of neurons tuned to particular stimulus attributes when these attributes are repeated (Grill-Spector and Malach 2001; Naccache and Dehaene 2001). This provides a useful method for probing the critical properties to which neurons respond, even when the different populations overlap in the same cortical region, since adaptation should occur for a repeated stimulus if the same neuronal population represents this stimulus across various appearances; whereas a lack of adaptation for a given stimulus repeated with a different appearance indicates the recruitment of a new population of neurons. Several studies found repetition-priming decreases in the FFA when faces were repeated but most have used the same photograph (Grill-Spector et al. 1999; Gauthier et al. 2000; Henson et al. 2000; Huettel and McCarthy 2001; Henson et al. 2002) or the same view with different renderings (George et al. 1999; Vuilleumier et al. 2003a; Eger et al. 2004).

In a first study (Pourtois et al. 2005b), unfamiliar faces were first shown in front-views or three-quarter views, and later repeated either with the same view (using different photographic shot) or with a different view. We found that the

FFA in both hemispheres showed view-sensitive repetition effects, with only a partial generalization from three-quarter to front views (Fig. 5). This indicates that face representation in the FFA is not view-invariant, and does not form a truly abstract and three-dimensional trace of faces after a single encounter. However, the asymmetrical pattern of repetition-priming effects (with some adaptation from three-quarter to front views but not vice versa) suggests that three-quarter views may provide more critical features to derive another view later, or provide better tridimensional cues relative to incomplete or inaccurate information in front-views. By contrast, we found that more medial regions in fusiform cortex showed repetition effects across all types of viewpoint changes, but these regions were outside face-selective areas and may contribute to higher-level processing stages related to associative processes related to semantic information or more abstract person-identity representations. Moreover, this generalization across viewpoints arose selectively in the left hemisphere. This hemispheric asymmetry might be consistent with other results showing that view-invariant priming effects for man-made objects were also selectively present in the left but not right anterior fusiform cortex (Vuilleumier et al. 2002).

A second fMRI study (Pourtois et al. 2005c) has recently confirmed that representation of faces in the FFA does not generalize across different views of the same identity, now using faces from both unfamiliar and famous people. We reasoned that famous faces would be more likely to give rise to a robust view-invariance in long-term representations as compared with unfamiliar faces viewed only once as in our previous study (Pourtois et al. 2005b). In this new experiment (Pourtois et al. 2005c), each individual face identity was first shown in a given view and then repeated in a different view after a varying delay (counterbalanced across subjects). Again, the FFA showed priming effects only when faces were repeated with the same view. There was no priming whatsoever in the FFA when the same face identity was repeated from one view to another, even for faces of famous people or actors that have repeatedly been seen under different appearance. All repetition effects for these well-known faces arose in left temporal and frontal cortex only, suggesting that they implicated more semantic information about person-identity rather than abstract visual representation of faces (Rhodes 1985; Damasio et al. 1990; Vuilleumier et al. 2003b).

This study also showed that a region in the medial fusiform gyrus, outside the FFA, showed some priming-realated decreases when unfamiliar faces were repeated with a slightly different viewpoint but still a similar appearance (Fig. 6). Unlike the previous study, this medial fusiform region was now found in the right but not left hemisphere. Thus, our results point to distinct subregions within fusiform cortex that may show a different sensitivity to viewpoint or visual similarity.

Taken together, our data do not support the hypothesis that the FFA may hold "face recognition units" representing faces in a view-independent format. Rather, face identity appears to be coded in a view-sensitive manner in the FFA, but it can generalize across different image renderings when these show the same viewpoint. Thus, memory traces of a given face identity might be represented in more distributed networks linking visual cortex with other distant brain areas

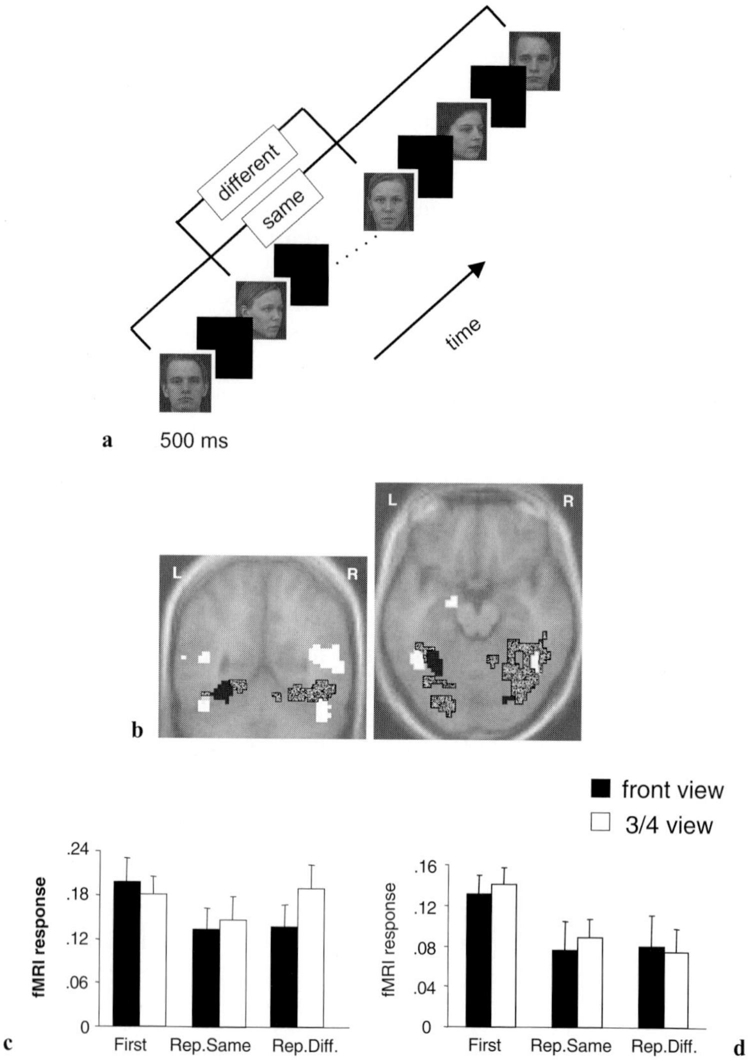

FIG. 5. **a** Stimuli used to test for repetition-priming effects when face identity is repeated either with the same viewpoint (*front-view* or *three-quarter*) or with a different viewpoint (e.g., first seen in front-view and later repeated in three-quarter, or vice versa). **b** Activation pattern across the different experimental conditions, overlaid on the mean anatomical scan of participants. White-colored areas show brain regions with face-selective responses, including FFA, STS, and amygdala. Gray-speckled areas show repetition-priming effects for faces repeated with the same view condition, involving extensive bilateral ventral temporal regions including FFA on both sides. Black-colored areas show repetition-priming effects for faces repeated with a different view, relative to faces seen for the fist time, involving the left medial fusiform cortex outside the FFA. Average parameter estimates of activity (±SE) are plotted for (**c**) the right FFA (red area) and (**d**) left medial fusiform cortex (blue area)

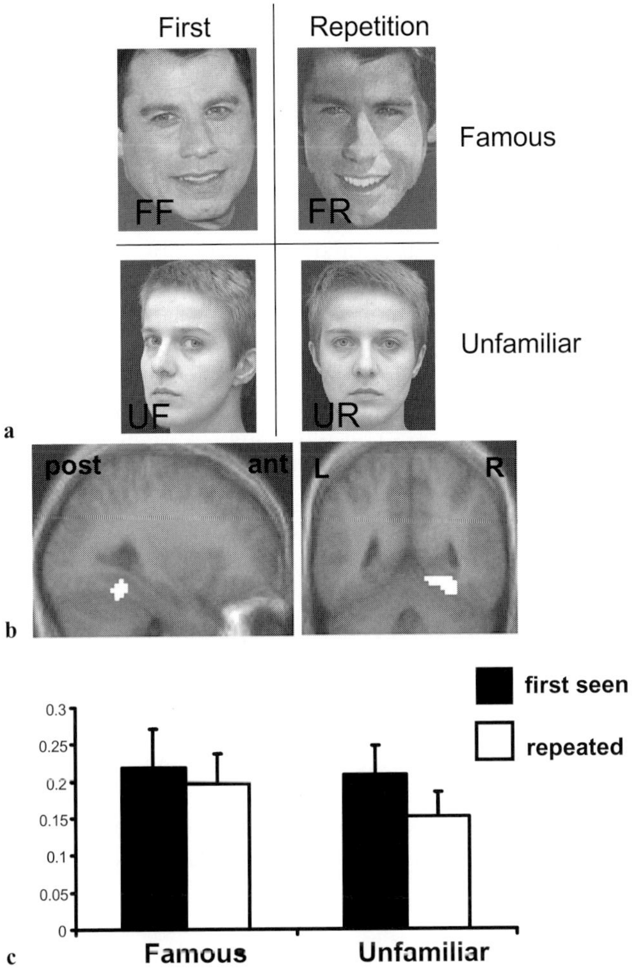

FIG. 6. **a** Stimuli used to test for repetition-priming effects when identity is repeated with the same or different viewpoint for either well-known or unknown faces. **b** Repetition-priming effects arose in a medial region of right fusiform cortex when identity was repeated across different views, but only for unknown faces which were visually more similar to each other, as compared to different views of famous faces which were visually more different. **c** Average parameter estimates of activity (±SE) in right fusiform cortex

(Bruce 1982; Damasio et al. 1990; Vuilleumier et al. 2003b), rather than being supported by in a single neuronal populations in a single brain area.

6 Conclusions

Recent brain imaging studies have highlighted the distributed and interactive nature of face perception in the human brain (Haxby et al. 2000, p. 256). The present chapter has focused on the processing of two major facial attributes (identity and expression) in the main brain regions associated with face recognition, i.e., the fusiform cortex (FFA), as well as the amygdala. Our findings reveal that although the FFA is critically implicated in face identity processing, repetition-priming effects may arise when the same face is seen across different picture formats but not when the same identity is seen across different viewpoint, suggesting that representations of faces in the FFA are not view-invariant and do not maintain a fully abstract 3D memory trace for previously encountered faces, even when these are from well-known people. In addition, face processing in the FFA is not totally independent of emotional expression, as predicted by traditional cognitive models proposing a strict segregation between processing pathways for expression and those for identity. However, emotional effects in the FFA are essentially generated by amygdala feedback on extrastriate cortex, which may arise during a second stage of processing after a first sweep of coarse visual inputs into the visual system. Future research still needs to elucidate the nature of visual information and computations taking place in different brain regions, and their dynamic interactions over time. Despite our impression that faces can be recognized effortlessly in a single glance, face recognition clearly involves more than a single brain process triggered in a single instant.

Acknowledgements. This work is supported by the Swiss National Science Fund (grant # 632.065935) and the Swiss National Center for Competence in Research in Affective Sciences.

References

Adolphs R, Gosselin F, Buchanan TW, Tranel D, Schyns P, Damasio AR (2005) A mechanism for impaired fear recognition after amygdala damage. Nature 433:68–72

Amaral DG, Behniea H, Kelly JL (2003) Topographic organization of projections from the amygdala to the visual cortex in the macaque monkey. Neuroscience 118:1099–1120

Bentley P, Vuilleumier P, Thiel CM, Driver J, Dolan RJ (2003) Cholinergic enhancement modulates neural correlates of selective attention and emotional processing. Neuroimage 20:58–70

Biederman I, Bar M (2000) Differing views on views: response to Hayward and Tarr (2000). Vision Res 40:3901–3905

Biederman I, Kalocsai P (1997) Neurocomputational bases of object and face recognition. Philos Trans R Soc Lond B Biol Sci 352:1203–1219

Bruce V (1982) Changing faces: visual and non-visual coding processes in face recognition. Br J Psychol 73:105–116

Bruce V, Young AW (1986) Understanding face recognition. Br J Psychol 77:305–327

Calder AJ, Young AW, Keane J, Dean M (2000) Configural information in facial expression. J Exp Psychol Hum Percept Perform 26:527–551

Damasio AR, Tranel D, Damasio H (1990) Face agnosia and the neural substrates of memory. Ann Rev Neurosci 13:89–109

Eastwood JD, Smilek D, Merikle PM (2001) Differential attentional guidance by unattended faces expressing positive and negative emotion. Percept Psychophys 63:1004–1013

Eger E, Schyns PG, Kleinschmidt A (2004) Scale invariant adaptation in fusiform face-responsive regions. Neuroimage 22:232–242

Fox E (2002) Processing of emotional facial expressions: the role of anxiety and awareness. Cogn Affect Behav Neurosci 2:52–63

Fox E, Lester V, Russo R, Bowles RJ, Pichler A, Dutton K (2000) Facial expressions of emotion: are angry faces detected more efficiently? Cogn Emot 14:61–92

Gauthier I, Tarr MJ, Moylan J, Skudlarski P, Gore JC, Anderson AW (2000) The fusiform "face area" is part of a network that processes faces at the individual level. J Cogn Neurosci 12:495–504

George N, Dolan RJ, Fink GR, Baylis GC, Russell C, Driver J (1999) Contrast polarity and face recognition in the human fusiform gyrus. Nat Neurosci 2:574–580

Gobbini MI, Haxby JV (2007) Neural systems for recognition of familiar faces Neuropsychologia 45(1):32–41

Grandjean D, Sander D, Pourtois G, Schwartz S, Seghier ML, Scherer KR, Vuilleumier P (2005) The voices of wrath: brain responses to angry prosody in meaningless speech. Nat Neurosci 8:145–146

Grill-Spector K, Malach R (2001) fMR-adaptation: a tool for studying the functional properties of human cortical neurons. Acta Psychol (Amst) 107:293–321

Grill-Spector K, Kushnir T, Edelman S, Avidan G, Itzchak Y, Malach R (1999) Differential processing of objects under various viewing conditions in the human lateral occipital complex. Neuron 24:187–203

Grill-Spector K, Knouf N, Kanwisher N (2004) The fusiform face area subserves face perception, not generic within-category identification. Nat Neurosci 7:555–562

Grüsser OJ, Landis T (1991) Visual agnosia and other disturbances of visual perception and cognition. In: Cronly-Dillon J (Ed) Vision and visual dysfunction, Vol. 12. MacMillan, London, pp 218–239

Hasselmo ME, Rolls ET, Baylis GC (1989) The role of expression and identity in the face-selective responses of neurons in the temporal visual cortex of the monkey. Behav Brain Res 32:203–218

Haxby JV, Horwitz B, Ungerleider LG, Maisog JM, Pietrini P, Grady CL (1994) The functional organization of human extrastriate cortex: a PET-rCBF study of selective attention to faces and locations. J Neurosci 14:6336–6353

Haxby JV, Hoffman EA, Gobbini MI (2000) The distributed human neural system for face perception. Trends Cogn Neurosci 4:223–232

Henson R, Shallice T, Dolan R (2000) Neuroimaging evidence for dissociable forms of repetition priming. Science 287:1269–1272

Henson RN, Shallice T, Gorno-Tempini ML, Dolan RJ (2002) Face repetition effects in implicit and explicit memory tests as measured by fMRI. Cereb Cortex 12:178–186

Huettel SA, McCarthy G (2001) Regional differences in the refractory period of the hemodynamic response: an event-related fMRI study. Neuroimage 14:967–976

Kanwisher N, McDermott J, Chun MM (1997) The fusiform face area: a module in human extrastriate cortex specialized for face perception. J Neurosci 17:4302–4311

Lane RD, Chua PM-L, Dolan RJ (1999) Common effects of emotional valence, arousal, and attention on neural activation during visual processing of pictures. Neuropsychologia 37:989–997

LeDoux JE (2000) Emotion circuits in the brain. Annu Rev Neurosci 23:155–184

McCarthy G, Puce A, Gore J, Allison T (1997) Face-specific processing in the human fusiform gyrus. J Cogn Neurosci 9:605–610

Merigan WH, Maunsell JH (1993) How parallel are the primate visual pathways? Annu Rev Neurosci 16:369–402

Mogg K, Bradley BP, Hallowell N (1994) Attentional bias to threat: roles of trait anxiety, stressful events, and awareness. Q J Exp Psychol A 47:841–864

Morris J, Friston KJ, Buchel C, Frith CD, Young AW, Calder AJ, Dolan RJ (1998a) A neuromodulatory role for the human amygdala in processing emotional facial expressions. Brain 121:47–57

Morris JS, Ohman A, Dolan RJ (1998b) Conscious and unconscious emotional learning in the human amygdala. Nature 393:470–474

Morris JS, Öhman A, Dolan RJ (1999) A subcortical pathway to the right amygdala mediating "unseen" fear. Proc Natl Acad Sci USA 96:1680–1685

Morris J, DeGelder B, Weiskrantz L, Dolan RJ (2001) Differential extrageniculostriate and amygdala responses to presentation of emotional faces in a cortically blind field. Brain 124:1241–1252

Morris JS, deBonis M, Dolan RJ (2002) Human amygdala responses to fearful eyes. Neuroimage 17:214–222

Mourao-Miranda J, Volchan E, Moll J, de Oliveira-Souza R, Oliveira L, Bramati I, Gattass R, Pessoa L (2003) Contributions of stimulus valence and arousal to visual activation during emotional perception. Neuroimage 20:1955–1963

Naccache L, Dehaene S (2001) The priming method: imaging unconscious repetition priming reveals an abstract representation of number in the parietal lobes. Cereb Cortex 11:966–974

Pasley BN, Mayes LC, Schultz RT (2004) Subcortical discrimination of unperceived objects during binocular rivalry. Neuron 42:163–172

Pegna AJ, Khateb A, Lazeyras F, Seghier ML (2005) Discriminating emotional faces without primary visual cortices involves the right amygdala. Nat Neurosci 8:24–25

Pessoa L (2005) To what extent are emotional visual stimuli processed without attention and awareness? Curr Opin Neurobiol 15:188–196

Pessoa L, Kastner S, Ungerleider LG (2002a) Attentional control of the processing of neural and emotional stimuli. Brain Res Cogn Brain Res 15:31–45

Pessoa L, McKenna M, Gutierrez E, Ungerleider LG (2002b) Neural processing of emotional faces requires attention. Proc Natl Acad Sci USA 99:11458–11463

Pourtois G, Grandjean D, Sander D, Vuilleumier P (2004) Electrophysiological correlates of rapid spatial orienting towards fearful faces. Cereb Cortex 14:619–633

Pourtois G, Dan ES, Grandjean D, Sander D, Vuilleumier P (2005a) Enhanced extrastriate visual response to bandpass spatial frequency filtered fearful faces: time course and topographic evoked-potentials mapping. Hum Brain Mapp 6:65–79
Pourtois G, Schwartz S, Seghier ML, Lazeyras F, Vuilleumier P (2005b) Portraits or people? View-sensitive and view-invariant memories of face identity in the human visual cortex. J Cogn Neurosci 17:1043–1057
Pourtois G, Schwartz S, Seghier ML, Lazeyras F, Vuilleumier P (2005c) View-independent coding of face identity in frontal and temporal cortices is modulated by familiarity: an event-related fMRI study. Neuroimage 24:1214–1224
Rhodes G (1985) Lateralized processes in face recognition. Br J Psychol 76:249–271
Sabatinelli D, Bradley MM, Fitzsimmons JR, Lang PJ (2005) Parallel amygdala and inferotemporal activation reflect emotional intensity and fear relevance. Neuroimage 24:1265–1270
Sahraie A, Weiskrantz L, Trevethan CT, Cruce R, Murray AD (2002) Psychophysical and pupillometric study of spatial channels of visual processing in blindsight. Exp Brain Res 143:249–256
Schyns PG, Oliva A (1999) Dr. Angry and Mr. Smile: when categorization flexibly modifies the perception of faces in rapid visual presentations. Cognition 69:243–265
Sergent J, Ohta S, Macdonald B (1992) Functional neuroanatomy of face and object processing: a positron emission tomography study. Brain 155:15–36
Sugase Y, Yamane S, Ueno S, Kawano K (1999) Global and fine information coded by single neurons in the temporal visual cortex. Nature 400:869–873
Surguladze SA, Brammer MJ, Young AW, Andrew C, Travis MJ, Williams SC, Phillips ML (2003) A preferential increase in the extrastriate response to signals of danger. Neuroimage 19:1317–1328
Tarr MJ, Bulthoff HH (1998) Image-based object recognition in man, monkey and machine. Cognition 67:1–20
Vuilleumier P (2005) How brains beware: neural systems for emotional attention. Trends Cogn Sci 9(12):585–594
Vuilleumier P, Schwartz S (2001a) Beware and be aware: capture of attention by fear-relevant stimuli in patients with unilateral neglect. Neuroreport 12:1119–1122
Vuilleumier P, Schwartz S (2001b) Emotional facial expressions capture attention. Neurology 56:153–158
Vuilleumier P, Armony JL, Driver J, Dolan RJ (2001) Effects of attention and emotion on face processing in the human brain: an event-related fMRI study. Neuron 30:829–841
Vuilleumier P, Henson R, Driver J, Dolan RJ (2002) Multiple levels of visual object constancy revealed by event-related fMRI of repetition priming. Nat Neurosci 5:491–499
Vuilleumier P, Armony JL, Driver J, Dolan RJ (2003a) Distinct spatial frequency sensitivities for processing faces and emotional expressions. Nat Neurosci 6:624–631
Vuilleumier P, Mohr C, Wenzel C, Valenza N, Landis T (2003b) Hyperfamiliarity for unknown faces after left lateral temporo-occipital venous infarction: a double dissociation with prosopagnosia. Brain 126:889–907
Vuilleumier P, Richardson M, Armony J, Driver J, Dolan RJ (2004) Distant influences of amygdala lesion on visual cortical activation during emotional face processing. Nat Neurosci 7:1271–1278
Whalen PJ, Rauch SL, Etcoff NL, McInerney SC, Lee MB, Jenike MA (1998) Masked presentations of emotional facial expressions modulate amygdala activity without explicit knowledge. J Neurosci 18:480–487

Whalen PJ, Kagan J, Cook RG, Davis FC, Kim H, Polis S, McLaren DG, Somerville LH, McLean AA, Maxwell JS, Johnstone T (2004) Human amygdala responsivity to masked fearful eye whites. Science 306:2061

Williams MA, Morris AP, McGlone F, Abbott DF, Mattingley JB (2004) Amygdala responses to fearful and happy facial expressions under conditions of binocular suppression. J Neurosci 24:2898–2904

Winston JS, Vuilleumier P, Dolan RJ (2003) Effects of low-spatial frequency components of fearful faces on fusiform cortex activity. Curr Biol 13:1824–1829

Wojciulik E, Kanwisher N, Driver J (1998) Covert visual attention modulates face-specific activity in the human fusiform gyrus: fMRI study. J Neurophysiol 79:1574–1578

Part II
Attention

9
Object Recognition: Attention and Dual Routes

Volker Thoma[1] and Jules Davidoff[2]

1 History of Hybrid Models

1.1 Introduction

The human capacity for visual object recognition is characterized by a number of properties that are jointly very challenging to explain. Recognition performance is highly sensitive to variations in viewpoint such as rotations in the picture plane (e.g., Murray 1995, 1998; Jolicoeur 1985) and to some rotations in depth (e.g., Hayward 1998; Lawson and Humphreys 1996, 1998) but invariant with the location of the image in the visual field (Biederman and Cooper 1991; Stankiewicz and Hummel 2002), the size of the image (Biederman and Cooper 1992; Stankiewicz and Hummel 2002), left-right (i.e., mirror) reflection (Biederman and Cooper 1991; Davidoff and Warrington 2001), and some rotations in depth (Biederman and Gerhardstein 1993). Second, object recognition is remarkably robust to variations in shape (Davidoff and Warrington 1999; Hummel 2001). For example, people spontaneously name the picture of a Collie or a Pomeranian both as simply a "dog" – a phenomenon termed "basic level" categorisation (Rosch et al. 1976).

Theorists traditionally struggle to account for these properties. In so called view-based theories (e.g., Olshausen et al. 1993; Poggio and Edelman 1990) representations mediating object recognition are usually based on metric templates derived from learned views. Although more recent accounts allow for combinations of template fragments (e.g., Edelman and Intrator 2003), the object features in view-based representations are fixed to certain locations in the image. Therefore, these accounts can readily explain effects of view-dependency in object recognition. In contrast, so-called structural description theories assume that the visual system extracts a more abstract representation from the 2D image on the

[1]School of Psychology, University of East London, Romford Road, London E15 4LZ, UK
[2]Psychology Department, Goldsmiths University of London, Lewisham Way, London SE14 6NW, UK

retina by encoding an object's constituent parts and their spatial relations (e.g., Biederman 1987; Hummel and Biederman 1992). Such a description is unaffected by many view-changes (such as changes in size, left-right reflection) and it also applies to many different exemplars of an object, permitting generalisation over metric variations of shapes (see Hummel 2001).

1.2 View Specific vs *Abstract Representations*

Not surprisingly, theorists have for some time sought to explain object recognition phenomena by integrating two qualitatively different types of representations. We will call these accounts hybrid models. For example, Posner and his colleagues (Posner 1969; Posner and Keele 1967) found an advantage for the sequential matching of identical letters in comparison with the matching of letters with the same name but differing case. However, this advantage was found only with short interstimulus intervals. These results were confirmed by other researchers with more realistic stimuli (Bartram 1976; Ellis et al. 1989; Lawson and Humphreys 1996) and were taken as evidence for the existence of a rapid, stimulus-specific representation and a more durable, abstract representation that generalises over variations in shape.

There is also neuropsychological evidence in support of representations that are either view-specific or more abstract. Warrington and her associates (Warrington and James 1988; Warrington and Taylor 1978) asked brain-damaged patients to recognize objects from canonical or non-canonical views. Observers with damage to the right posterior areas of the brain were particularly poor at non-canonical object recognition; therefore, Warrington and Taylor (1978) proposed that visual object recognition involves in two main stages. In the first stage, perceptual object constancy is achieved, relying heavily on right hemisphere processing. The second stage involves semantic categorisation, which taps primarily left hemisphere processing. Damage to the right hemisphere would therefore impair object constancy, so that only objects in highly familiar (canonical) views are recognisable (Warrington and James 1988). There are more recent accounts based on such hemispheric differences in which an abstract-category recognition system is assumed to be dominant in the left brain hemisphere whereas a specific-exemplar subsystem is thought to be working more effectively in the right hemisphere (Marsolek 1999).

Somewhat different representations working in two parallel pathways were proposed by Humphreys and Riddoch (1984). Their patients were shown 3 photographs of objects. The task was to match two different views of a target object by discriminating the object from a visually similar distracter object. Four of their patients with right-hemisphere damage only showed impairment in this task when the principal axis of the target object was foreshortened in one of the photographs. In contrast, a fifth patient (with damage to the left hemisphere) showed impaired matching only when the saliency of the target object's main distinctive feature was reduced, but foreshortening of the principal axis did not affect his performance. According to Humphreys and Riddoch (1984), this double dissociation indicates that two functionally independent routes are responsible for achieving object constancy. One route processes an object's local distinctive features whereas the second route encodes the object's structure relative to the frame of its principal axis.

One particular shortcoming of these early hybrid accounts discussed above is their lack of specification. In particular, it is not clear under what conditions the different representations are tapped separately or in combination. One type of attempt to clarify those conditions is to invoke process differences such as mental rotation (Jolicoeur 1990; Corballis 1988) or holistic *vs* analytic processing (Farah 1990, 1991). These will not be dealt with here but for a critical review, see Humphreys and Rumiati (1998) and Lawson (1999).

1.3 Representation Use according to Task-Demands

Tarr and Bulthoff (1995) suggest that human object recognition can be thought of as a continuum between pure exemplar-specific discriminations and categorical discriminations. According to this line of thinking, extreme cases of within-class discriminations allow for recognition exclusively achieved by viewpoint-dependent mechanisms. When objects are to be distinguished in broad categorical classes recognition of objects may be exclusively achieved by viewpoint-invariant mechanisms. Shape discriminations usually fall within the extremes of the continuum and recognition is mediated by viewpoint-dependent and viewpoint-independent mechanisms according to the nature of the task, the similarity and familiarity of the stimuli, and other context conditions. Although this account seems intuitive, its predictions are rather general and the experimental evidence is somewhat unclear (Murray 1998; Hayward and Williams 2000).

2 A Hybrid Model of Object Recognition and Attention

2.1 The Hummel Model

Most of the previous hybrid accounts incorporate representations that have properties similar to structural descriptions (e.g., Hummel and Biederman 1992) or view-like representations (e.g., Olshausen et al. 1993). However, which type of representation is employed may depend on attention (Hummel and Biederman 1992). The next section will describe a hybrid account of object recognition that specifies how visual attention affects the representation of object shape.

The fact that both structural descriptions and view-based representations of shape can account for some, but not all of the properties of object recognition led Hummel (Hummel and Stankiewicz 1996; Hummel 2001) to propose that objects are recognized based on a hybrid representation of shape, consisting of

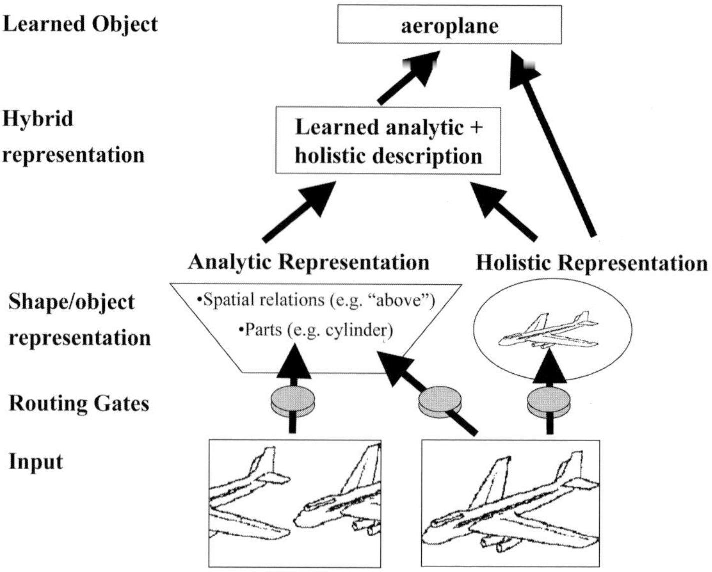

FIG. 1. A simple sketch of the architecture of JIM.3 (adapted from Hummel 2001). Units in the input layers of the model are activated by the contours from an object's line drawing. Routing gates propagate the output to units with two representational components: The independent units represent the shape attributes of an object's geons, and the units in the holistic map represent shape attributes of surfaces. The activation patterns of both components are learned individually, then summed in a higher layer over time. Units in the uppermost layer code object identity

a holistic (i.e., "view"-like) representation as well as an analytic representation (i.e., a structural description) of shape (Fig. 1).

Given a 2D image (such as a line-drawing) of an object, the Hummel model (JIM.3) generates both an analytic and a view-based representation. The analytic representation codes an object's shape in terms of the object's parts and their categorical interrelations. This representation has the properties of structural description (Biederman 1987) and is largely robust to many variations in viewpoint (such as translation, changes in scale, left-right reflection and some rotations in depth) but it is sensitive to rotations in the picture plane (see Hummel and Biederman 1992). The analytic representation allows generalization to novel views and to novel exemplars of known categories. However, it requires processing time and visual attention to be able to represent parts and spatial relations independently of each other (Hummel and Biederman 1992; Hummel 2001).

The holistic representation, in contrast, does not specify parts of an object or their categorical spatial relations. Instead, object parts are represented in terms of their topological positions in a 2-D coordinate system (see Hummel 2001). Since the holistic representation does not require attention for binding parts to their spatial relations, it can be generated rapidly and automatically. The representation formed on the holistic map is sensitive to left-right reflections as well as to rotations in the picture plane and in depth because the units representing object surfaces are spatially separated. However, the holistic representation is invariant with translation and scale.

2.2 Previous Tests of the Hummel Model

Stankiewicz et al. (1998) tested the predictions of the hybrid analytic/holistic model regarding changes in viewpoint using an object naming task with paired prime/probe trials. A prime trial consisted of a fixation cross followed by a box to the left or right of fixation, which served as an attentional cue (see Fig. 2 for

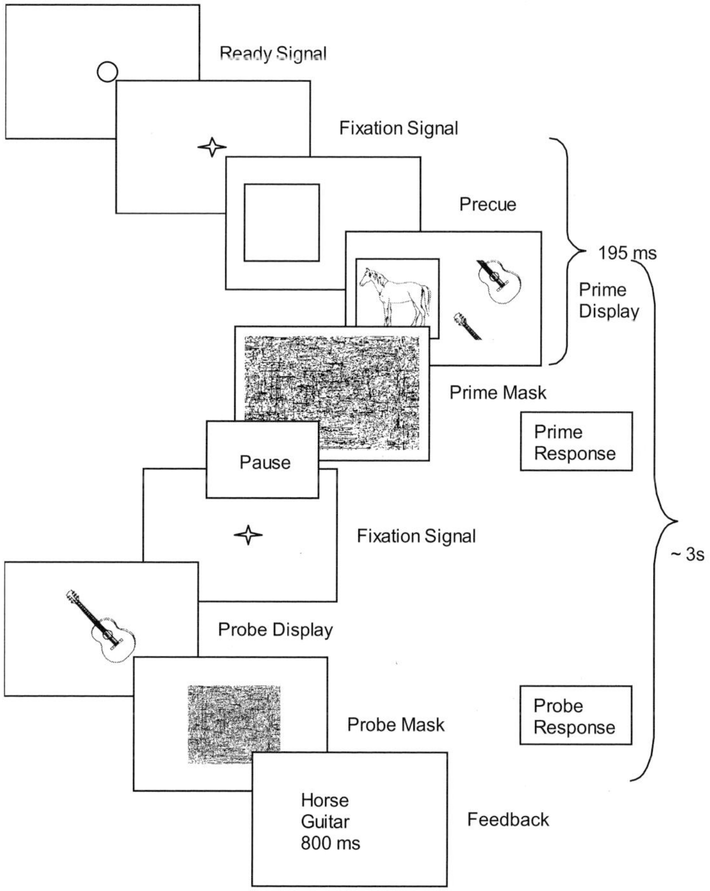

Fig. 2. Sequence of displays in a typical short-term priming paradigm (here an example from Experiment 1)

a similar paradigm used in Thoma et al. 2004). This was followed by two line drawings of common objects, one of which appeared inside the cueing box, and the other appeared on the other side of fixation. The participant's task was to immediately name only the cued image (the attended prime) and not respond to the other image (the ignored prime). The entire prime trial (from cueing box to mask) lasted only 195 ms, which is too brief to permit a saccade away from fixation. Each prime display was masked and after 2 seconds followed by a probe display containing a single image of an object at fixation. Again, the task was to name the object which was either the same object as the attended prime, the same object as the ignored prime, or an object the participant had not previously seen in the experiment (an unprimed probe, which served as a baseline to measure priming). Images of repeated objects (i.e., other than unprimed probes) were either identical to the corresponding primes, or were left-right reflections of them. Priming was measured as the difference in latencies between repeated (previously attended or ignored) and unrepeated (unprimed) probe images. The results showed that attended prime images reliably primed both themselves and their left-right reflections. However, ignored prime images only primed themselves in the same view. Moreover, the effects of attention (attended vs. ignored) and reflection (identical images vs. left-right reflections) were strictly additive: The priming advantage for same view prime-probe trials was equivalent in both attended and unattended conditions (about 50 ms). The fact that attention and reflection had additive effects on priming provides strong support for the independence of the holistic and structured representations of shape in the hybrid model. A holistic representation contributes to priming in a strictly view-dependent way and is independent of attention, whereas an analytic representation contributes to priming regardless of the view but depends on attention. Stankiewicz and Hummel (2002) tested the hybrid model's predictions concerning changes in position and scale using a similar paradigm as in Stankiewicz et al. (1998). As predicted, priming for attended and ignored objects was not affected by view changes such as translation and scaling (i.e., changes in position and size).

2.3 Testing Configural Distortions in the Hybrid Model

Here, we report 8 further experiments that examine aspects of the Hummel model using a priming paradigm similar to that employed by Stankiewicz et al. (1998). The findings of Stankiewicz and colleagues are clearly consistent with the hybrid model, but they cannot provide a direct test for the model's primary theoretical assertion – that the object shape is represented in a hybrid analytic and holistic fashion. To test the assumption of truly analytic representations underlying object recognition, we employed images that would not resemble any holistic representations. Whereas analytic representations of shape should be necessarily robust to configural distortions – such as scrambling of component parts – a holistic representation should be very sensitive to such image variations. Consider the manipulation of splitting an image down the middle and moving the left half of the image to the right-hand side (Fig. 1). A holistic representation of the intact aeroplane (e.g., stored as a view as in a typical image-based model; e.g., Poggio and Edelman 1990) would be matched, in its entirety, against an object's image to determine the degree of fit between the image and the holistic representation (i.e., view) in memory.

According to this holistic measure of similarity, the intact and split images of the aeroplane are very much different. However, a structural representation could compensate for this manipulation as long as the shapes of the object's parts are recoverable from the information presented in each half of the image (Biederman 1987; Hummel and Biederman 1992). In the split image, the front of the aeroplane is not connected to the back, yet the two halves retain enough structural information to allow the identification of the object.

Experiments 1–3 are from Thoma et al. (2004) and were designed to directly test the central theoretical assertion of the hybrid model that the representation of an attended image is analytic and holistic whereas the representation of an ignored image is only holistic. Experiment 1 investigated the role of attention in priming for split and intact object images. Participants named objects in pairs of prime-probe trials (as in Stankiewicz et al. 1998). Half of the prime images were presented intact, and half were split either horizontally or vertically, as illustrated in Figure 2. The factors of attention (attended *vs* ignored image) and image type (intact vs. split) were crossed orthogonally. The probe image was always intact and corresponded either to the attended prime, the ignored prime, or it was an image the observer had not previously seen in the experiment (which served as a baseline).

As predicted by the hybrid model, split images primed their intact counterparts only when the split images were attended, but both attended and ignored intact images primed their intact counterparts (see Fig. 3a). There was a reliable priming advantage for intact primes over split primes. Thus, the effects of attention (attended *vs* ignored) and configuration (intact vs. split) were strictly additive as in Stankiewicz et al. (1998).

Experiment 2 was designed to estimate what fraction of the priming observed in Experiment 1 was due to visual (as opposed to concept and/or name) priming. Images in the identical-image conditions of Experiment 1 (attended-intact, ignored-intact) were replaced with images of objects having the same basic-level name (e.g., "piano") as the corresponding probe object, but with a different shape (e.g., "grand piano" instead of "upright piano"). The results of Experiment 2 showed than an intact probe image was primed more (about 80 ms) by an attended split image of the same exemplar (e.g., a grand piano) than by an attended intact different exemplar of the same basic-level category (e.g., upright piano). Since in both cases participants responded with the same name in prime and probe trials, this difference indicates a strong visual component to the priming in the attended and ignored conditions. There was no priming for unattended primes (split or different exemplar), suggesting that all the priming observed in the unattended condition of Experiment 1 was specifically visual.

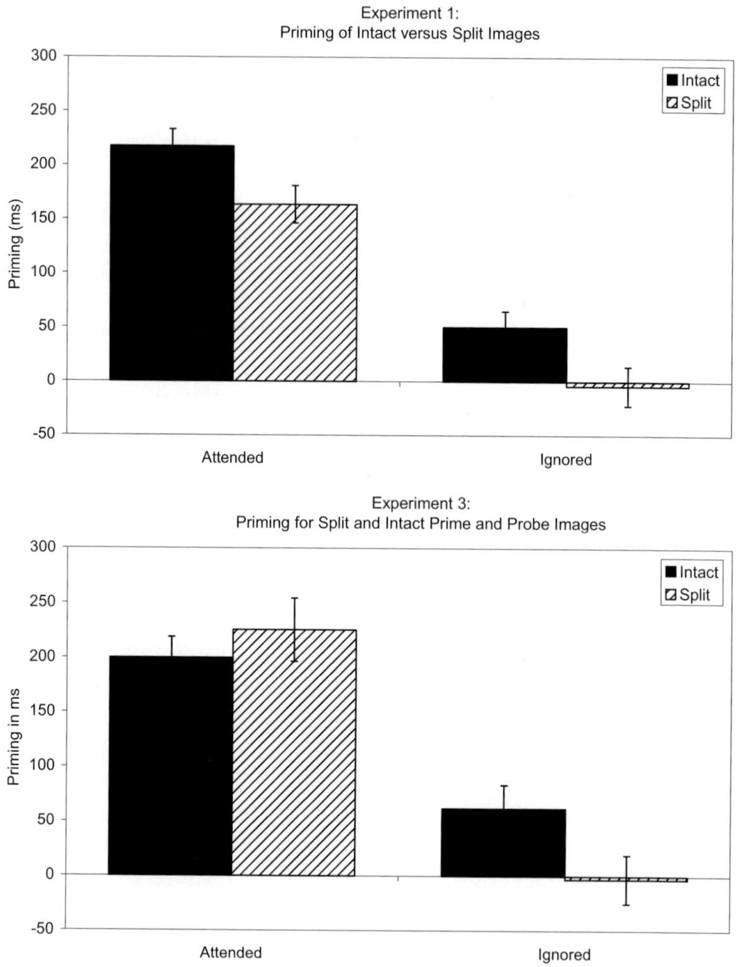

a b

FIG. 3a,b. Priming (baseline RT minus RT in each experimental condition) means (ms) and standard errors in **a** Experiment 1 for intact probe images (Thoma et al. 2004) as a function of whether the prime image was attended or ignored and intact or split (n = 42). **b** Priming means in ms and standard errors for Experiment 3 (Thoma et al. 2004) as a function of whether the prime object was attended or ignored and whether both prime and probe were split or intact

The priming patterns in Experiments 1 and 2 are predicted by the theory that the visual system generates holistic representations of ignored images and analytic representations of attended images (Hummel 2001; Hummel and Stankiewicz 1996). However, an alternative interpretation is that all the observed priming resides in early visual representations (i.e., rather than in representations responsible for object recognition, as assumed by the hybrid model). Identical images may simply prime one another more than non-identical images, and attended images prime one another more than unattended images. If this "early priming" account is correct, then the advantage for identical images over non-identical images and the advantage for attended images over unattended images could yield the effects found in Experiment 1. This interpretation is challenged by the results of Stankiewicz and Hummel (2002), who showed that priming for ignored images is invariant with translation and scale. Nevertheless, a third experiment was designed to establish whether the results of Experiments 1 and 2 reflect a reliance on holistic processing for ignored images, as predicted by the hybrid model. Applied to the current paradigm, the logic is as follows: If the results of the first two experiments reflect the role of holistic representations in the recognition of ignored images, and if these holistic representations are encoded in LTM in an intact (rather than split) format, then ignoring a split image on one occasion should not prime recognition of the very same image on a subsequent occasion. However, if the results are due to priming early visual features (in both the attended and ignored cases), then ignoring a split image on one trial should prime recognition of that (split) image on the subsequent trial. By contrast, both models would predict that attending to a split image on one trial should permit the encoding and, therefore, priming of that image.

The results of Experiment 3 showed that a split image primed itself when attended but not when ignored, whereas an intact image primed itself under both conditions (see Fig. 3b). Critically, in the ignored conditions priming was found only for intact images but not for repeated split images. This demonstrated that the lack of priming for ignored split images in Experiment 1 cannot be attributed to a general decrease of priming in response to split images. The priming pattern is predictable from the hybrid model and in contrast to the alternative hypothesis that would have predicted equal levels of priming under both ignored conditions.

The results reported by Thoma et al. (2004) strongly support the central theoretical tenet of the hybrid model of object recognition (Hummel 2001; Hummel and Stankiewicz 1996), that object recognition is based on a hybrid analytic + holistic representation of object shape. Attended intact, attended split and ignored intact images primed subsequent recognition of corresponding intact images, whereas ignored split images did not prime their intact counterparts. This pattern of effects is predicted by the hybrid account because attended images are represented both analytically and holistically, whereas ignored images are represented only holistically.

2.4 Plane Rotations

Object recognition is well-known to be sensitive to orientation in the picture plane (for a review, see Lawson 1999). The principal aim of Experiments 4 and 5 (Thoma, Davidoff, and Hummel 2007) was to test the hybrid model with picture plane rotations. A distinction between base (objects with a preferred upright position) and no-base objects (objects without a definite base) was made, which has previously been found to have importance for both behavioural (Vannucci and Viggiano 2000) and neuropsychological (Davidoff and Warrington 1999) investigations of object orientation. In a simple object naming study Thoma et al. (2007) confirmed the finding that objects with a definite base (e.g., a house) incurred increasing recognition performance costs when rotated, whereas no-base objects (e.g., hammer) were equally recognisable in all picture plane orientations. Subsequently, in Experiment 4, the pattern of priming effects observed for plane-rotated no-base objects clearly replicated the findings of Stankiewicz et al. (1998) with mirror-images and those of Thoma et al. (2004) with split images. Thus, the general notion of a hybrid model consisting of a holistic and analytic representation is supported by the fact that attended objects primed themselves in both the same view and the rotated view, whereas ignored objects only primed themselves in the same view (Hummel 2001).

In an attempt to test whether low-level early priming could have yielded these results, a replication of Experiment 3 was attempted using base-objects (e.g., a house). The relevant prime objects (attended or ignored) and the corresponding probe images were shown in the same orientation – both appeared in either an upright (familiar) or rotated (unfamiliar) view. The particular interest was in the ignored trials. Once more, Experiment 5 found a significant amount of priming in one condition (upright prime and identical probe image) and no priming in the other. Importantly, the lack of priming here was for ignored identical views that were unfamiliar (rotated view of base objects). Thus, the lack of priming in the ignored conditions for rotated no-base objects seen in Experiment 4 cannot be attributed to changes in early visual stimulation and cannot be trivially attributed to the amount of featural overlap between prime and target views. The priming pattern found in Experiment 5 (and previously with split objects, see Fig. 3b) is perhaps the most direct evidence that images of ignored objects achieve priming from access to stored familiar views. The data also fit previous findings that attention is necessary to establish view-independent representations (Murray 1995).

2.5 Depth Rotation

Experiments 6–8 (Thoma and Davidoff 2006) are concerned with depth rotations in the Hummel model. Just as with plane-rotations there are many documented effects of rotations in depth on recognition performance. Many researchers have shown view-dependent effects after depth rotations of familiar objects (e.g., Hayward 1998, Lawson and Humphreys 1996, 1998). However, Biederman and his colleagues (Biederman and Gerhardstein 1993) have obtained view-invariant effects after some rotations in depth that did not alter the visible part-structure of an object. The hybrid theory of object recognition may offer an explanation for mixed findings on depth-rotation effects.

Certain rotations in depth produce a mirror transformation of the image if the object is bilaterally symmetric. In Experiment 6, the findings of Stankiewicz et al. (1998) with mirror images were replicated with a new set of photorealistically rendered objects. Once more, the effects of attention and viewpoint were additive. Attended objects primed both themselves and their reflected versions, whereas ignored objects only primed themselves but not their mirror versions. Thus, the hybrid model may account for effects of depth rotations in which the part structure is not changed between views.

In contrast to mirror reflections, rotations in depth between study and test can affect the analytic representation because visible parts may be occluded or new parts may be revealed (Biederman and Gerhardstein 1993). Depth-rotations that differ from those akin to mirror-reflections should therefore provide an opportunity to further test the theory that two representations work in parallel because depth rotation may affect both representational components (analytic and holistic) instead of just one (holistic). The aim was to test whether depth-rotation involving part changes affects priming for attended objects (analytic plus holistic representation) more than for ignored objects (holistic representation only).

The logic underlying Experiment 7 comprises three parts: First, according to the hybrid model, all viewpoint changes (except translation and scaling) should affect the holistic component. Second, because the holistic representation works with and without attention, changes in viewpoint by depth-rotations should equally decrease the amount of priming in both attended and ignored conditions compared to priming in the identical viewpoint. Third, depth-rotations that affect the perceived part structure of the object should additionally reduce the amount of priming for attended images (because only then will the analytic representation be affected), but not for ignored images. In summary, if a part-based representation is involved for attended images but not for ignored ones, object rotations involving part changes should affect priming for attended images (holistic and analytic change) more than for ignored images (holistic change only).

In Experiment 7, objects were rotated in depth to produce an altered part-structure between views. To achieve a qualitative change in view orientation, objects were rotated in depth and depicted in two views. One was a complete side view (Fig. 4c) that would be primed by a more conventional view or vice versa (Fig. 4b). As a consequence, some parts of the object seen in one view (Fig. 4c) are not visible in the second view (e.g., the tail in Fig. 4b) and vice versa (e.g., legs in Fig. 4c). The effect of part-change was verified in a pilot study.

FIG. 4a–c. Three views of an example object as used in Thoma and Davidoff (2006). View **b** is rotated further away (90°) from view **a** than from view **c** (60°), but the object shares more visible parts with view **a**, because two of the legs are hidden in view **c** whereas a new part (the tail) appears

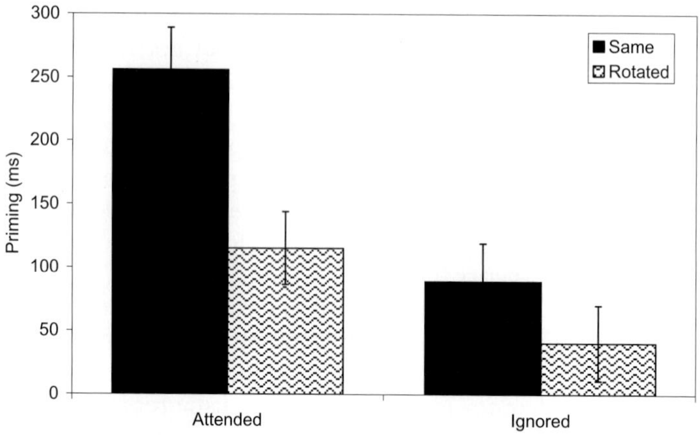

FIG. 5. Priming means (ms) and standard errors in Experiment 7 (Thoma and Davidoff 2006, Experiment 2) as a function of whether the object was attended or ignored in the prime display and whether the probe objects were presented in the same orientation or rotated in depth

The results of Experiment 7 replicated the previous findings of priming for attended images in the same view and in a changed (here: depth-rotated) orientation while ignored objects only primed themselves in the same view (see Fig. 5). Unlike previous tests of the hybrid model, the data show a unique interaction between attention and view-change: The difference between identical and depth-rotated views was significantly greater for attended than for ignored images.[1] This novel priming pattern is in line with the prediction of the hybrid model that depth-rotations may cause qualitative changes in analytic representations that depend on attention.

The data are not in line with view-based accounts. If attention plays a role in matching input with representations based on metric properties, one would expect enhanced priming effects for rotated objects in attended conditions relative to ignored conditions because attention would serve to aid the matching process (e.g., Olshausen et al. 1993). This was not the case here – the priming difference between rotated and identical view was greater for attended than for ignored objects.

As predicted from the hybrid model of Hummel (2001), viewpoint and attention produced additive effects of priming between qualitatively similar views, just as observed in Experiment 6. In Experiment 8, there was a greater degree of angular separation (90°) between the prime and probe view than in Experiment 7 (60°), yet the former view pairs (Fig. 4a,b) were rated by observers as more similar (in terms of visible parts) than the view pairs of Experiment 7 (Fig. 4b,c). Thus, the differences between the attended conditions of Experiments 7 and 8 confirm previous findings (Hayward 1998; Lawson 1999) that the amount of angular rotation (60° *vs* 90°) is not a reliable predictor of recognition performance as would be expected if object shape was represented only metrically.

The results for attended images also confirm that object recognition depends on whether the same parts are visible across views (Biederman and Gerhardstein 1993; Srinivas 1995). The hybrid's model general notion that object recognition across rotations in depth involves both an analytic and a holistic representation is also corroborated by Foster and Gilson (2002) who used novel 3-D objects that were to be discriminated in matching tasks either by a metric or a non-accidental (i.e., structural) property.

3 Multiple Representations in the Brain

The recent confirmation of the Hummel model from behavioural evidence finds support from neuroscience. Janssen et al. (2000) showed that neurons in the superior temporal sulcus were selective for three-dimensional shape whereas neurons in the lateral TE were generally unselective for 3D shape, though equally selective for 2D shape. Functional imaging studies (e.g., Vuilleumier et al. 2002) also support the notion that two types of object representations can be distinguished according to view-invariance in priming tasks. Vuilleumier et al. (2002) showed that repetition of images of common objects decreased activity (i.e., showed priming) in the left fusiform area independent of viewpoint (and size), whereas a viewpoint-dependent decrease in activation was found in the right fusiform area. Interestingly, the latter area was sensitive to changes in orientation but not in size – properties of the holistic component directly predicted by the hybrid model (Hummel 2001) and confirmed in behavioral studies (Stankiewicz and Hummel 2002).

As we have outlined in a previous section, numerous studies of patients with (limited) object agnosia indicate qualitatively different representations (Warrington and James 1988; Humphreys and Riddoch 1984). More recent evidence seems to corroborate the idea of multiple representations in the brain. Davidoff and Warrington (1999, 2001) studied patients who were extremely impaired at recognising object parts. Nevertheless, they were normal in naming intact objects though only when seen in familiar views. In terms of the hybrid model, the patients' holistic components seemed intact, allowing object recognition from familiar views, whereas analytic components were impaired preventing recognition of object parts or from unfamiliar views.

There is also neuropsychological evidence that attention may play a role in shape representation. Patients demonstrating unilateral neglect usually fail to respond to stimuli presented on the side contralateral to their lesion. Despite showing poor response to contralesional stimuli, there is evidence that these patients can nevertheless process semantic and shape properties in that field (Marshall and Halligan 1988; McGlinchey-Berroth et al. 1993). Recently, Forti and Humphreys (in press) have shown that the processing of shape information in the neglected hemifield depends on viewpoint as proposed by Stankiewicz et al. (1998) and seems qualitatively different from non-neglected stimuli. Similar findings come from studies on extinction, in which patients are able to detect ipsilesional stimuli presented alone but not when they are presented simultaneously with a stimulus on the contralesional side. Importantly, a recovery from extinction can be observed for global form information (Humphreys et al. 2000).

4 Conclusions

As we have seen, studies from different areas of cognitive science indicate the coexistence of multiple or hybrid representations of shape, resembling a distinction between holistic and analytic processing (Hummel 2001). This chapter has focused on the processing of shape in Hummel's model of object recognition because it is currently the most detailed model describing the role of attention in hybrid representation. Studies using traditional (rotation, reflection, scaling, translation, exemplar change) and novel (splitting) manipulations of object shape clearly confirmed the model's predictions regarding analytic/holistic representations. However, there are still many aspects of object recognition that are yet to be integrated into the model.

The hybrid model is largely based on a structural descriptive approach to object recognition (Hummel and Biederman 1992), which has been criticized in the past (e.g., Tarr and Bulthoff 1995; Edelman and Intrator 2003). For example, it is unclear how the model (and its predecessors) extracts axes of geons from 2D images. Another critique concerns the representation of irregular objects without obvious parts (such as a bush). One solution could be that in these cases, recognition relies more on the holistic component (Hummel 2003). A further way in which aspects of the Hummel model may be employed is to consider the role of time. For example, Zago et al. (2005) showed that visual priming for objects was maximal for an exposure time of 250 ms, then decreases. Therefore, they argued that certain aspects of an initial broad representation may be fine-tuned, becoming more stimulus specific.

In summary, it seems that attention is not necessary for object recognition but that the representations underlying object recognition differ according to whether an object is attended or not. An analytic representation is formed for attended objects and will be relatively robust to changes in view or configuration, except for part-changes. A holistic representation of an object is formed with and without attention allowing rapid recognition, but such a representation is very sensitive to any changes in view of global shape.

Note

1. The level of priming in Experiment 7 under all priming conditions was slightly higher than that in other experiments, and there was a slight trend toward positive priming for the ignored rotated prime. This may be due to the fact that the probe views were slightly less canonical (foreshortened) which produced longer identification times for baseline conditions (~50 ms compared to Experiment 6) and allowed more room for priming.

References

Bartram DJ (1976) Levels of coding in picture-picture comparison tasks. Mem Cognit 4:593–602

Biederman I (1987) Recognition-by-components: a theory of human image understanding. Psychol Rev 94:115–147

Biederman I, Cooper EE (1991) Evidence for complete translational and reflectional invariance in visual object priming. Perception 20:585–593

Biederman I, Cooper EE (1992) Size invariance in visual object priming. J Exp Psychol Hum Percept Perform 18:121–133

Biederman I, Gerhardstein PC (1993) Recognizing depth-rotated objects – evidence and conditions for 3-dimensional viewpoint invariance. J Exp Psychol Hum Percept Perform 19:1162–1182

Corballis MC (1988) Recognition of disoriented shapes. Psychol Rev 95:115–123

Davidoff J, Warrington EK (1999) The bare bones of object recognition: implications from a case of object recognition impairment. Neuropsychologia 37:279–292

Davidoff J, Warrington EK (2001) A particular difficulty in discriminating between mirror images. Neuropsychologia 39:1022–1036

Edelman S, Intrator N (2003) Towards structural systematicity in distributed statically bound visual representations. Cogn Sci 27:73–110

Ellis R, Allport DA, Humphreys GW, Collis J (1989) Varieties of object constancy. Q J Exp Psychol A 41A:775–796

Farah MJ (1990) Visual agnosia: disorders of object recognition and what they tell us about normal vision. MIT Press/Bradford Books, Cambridge MA

Farah MJ (1991) Patterns of co-occurrence among the associative agnosias: implications for visual object representation. Cognit Neuropsychol 8:1–19

Forti S, Humphreys GW (in press) Representation of unseen objects in visual neglect: effects of view and object identity. Cognit Neuropsychol (in press)

Foster DH, Gilson SJ (2002) Recognizing novel three-dimensional objects by summing signals from parts and views. P Roy Soc Lond B Bio 269:1939–1947

Hayward WG (1998) Effects of outline shape in object recognition. J Exp Psychol Hum Percept Perform 24:427–440

Hayward WG, Williams P (2000) Viewpoint dependence and object discriminability. Psychol Sci 11:7–12

Hummel JE (2001) Complementary solutions to the binding problem in vision: implications for shape perception and object recognition. Vis Cogn 8:489–517

Hummel JE (2003) The complementary properties of holistic and analytic representations of object shape. In: Rhodes G, Peterson M (Eds) Analytic and holistic processes in the perception of faces objects and scenes. Greenwood, Westport CT, pp 212–234

Hummel JE, Biederman I (1992) Dynamic binding in a neural network for shape-recognition. Psychol Rev 99:480–517

Hummel JE, Stankiewicz BJ (1996) An architecture for rapid hierarchical structural description. In: Inui T, McClelland J (Eds) Attention and performance XVI: information integration in perception and communication. MIT Press, Cambridge MA, pp 93–121

Humphreys GW, Riddoch MJ (1984) Routes to object constancy – implications from neurological impairments of object constancy. Q J Exp Psychol A 36:385–415

Humphreys GW, Rumiati RI (1998) Agnosia without prosopagnosia or alexia: evidence for stored visual memories specific to objects. Cognit Neuropsychol 15:243–277

Humphreys GW, Cinel C, Wolfe JM, Olson A, Klempen N (2000) Fractionating the binding process: neuropsychological evidence distinguishing binding of form from binding of surface features. Vision Res 40:1569–1596

Janssen P, Vogels R, Orban GA (2000) Selectivity for 3D shape that reveals distinct areas within macaque inferior temporal cortex. Science 288:2054–2056

Jolicoeur P (1985) The time to name disoriented natural objects. Mem Cognit 13:289–303

Jolicoeur P (1990) Identification of disoriented objects: a dual systems theory. Mind Lang 5:387–410

Lawson R (1999) Achieving visual object constancy across plane rotation and depth rotation. Acta Psychol 102:221–245

Lawson R, Humphreys GW (1996) View specificity in object processing: evidence from picture matching. J Exp Psychol Hum Percept Perform 22:395–416

Lawson R, Humphreys GW (1998) View-specific effects of depth rotation and foreshortening on the initial recognition and priming of familiar objects. Percept Psychophys 60:1052–1066

Marshall JC, Halligan PW (1988) Blindsight and insight in visuo-spatial neglect. Nature 336:766–767

Marsolek CJ (1999) Dissociable neural subsystems underlie abstract and specific object recognition. Psychol Sci 10:111–118

McGlinchey-Berroth R, Milberg W, Verfaellie M, Alexander M, Kilduff PT (1993) Semantic processing in the neglected visual field: evidence from a lexical decision task. Cognit Neuropsychol 10:79–108

Murray JE (1995) The role of attention in the shift from orientation-dependent to orientation-invariant identification of disoriented objects. Mem Cognit 23:49–58

Murray JE (1998) Is entry-level recognition viewpoint invariant or viewpoint dependent? Psychon B Rev 5:300–304

Olshausen B, Anderson C, Van Essen D (1993) A neurobiological model of visual attention and invariant pattern recognition based on dynamic routing of information. J Neurosci 13:4700–4719

Poggio T, Edelman S (1990) A network that learns to recognize 3-dimensional objects. Nature 343:263–266

Posner MI (1969) Abstraction and the process of recognition. In: Spence JT, Bower G (Eds) The psychology of learning and motivation. Academic Press, New York, pp 43–100

Posner MI, Keele SW (1967) Decay of visual information from a single letter. Science 158:137–139

Rosch E, Mervis CB, Gray WD, Boyes-Braem P (1976) Basic objects in natural categories. Cognit Psychol 8:382–439

Srinivas K (1995) Representation of rotated objects in explicit and implicit memory. J Exp Psychol-Hum L 21:1019–1036

Stankiewicz BJ, Hummel JE (2002) Automatic priming for translation- and scale-invariant representations of object shape. Vis Cogn 9:719–739

Stankiewicz BJ, Hummel JE, Cooper EE (1998) The role of attention in priming for left-right reflections of object images: evidence for a dual representation of object shape. J Exp Psychol Hum Percept Perform 24:732–744

Tarr MJ, Bulthoff HH (1995) Is human object recognition better described by geon structural descriptions or by multiple views – comment on Biederman and Gerhardstein (1993). J Exp Psychol Hum Percept Perform 21:1494–1505

Thoma V, Davidoff J (2006) Priming for depth-rotated objects depends on attention and part changes. Exp Psychol 53:31–47
Thoma V, Hummel JE, Davidoff J (2004) Evidence for holistic representation of ignored images and analytic representation of attended images. J Exp Psychol Hum Percept Perform 30:257–267
Thoma V, Davidoff J, Hummel JE (2007) Priming of plane-rotated objects depends on attention and view familiarity. Vis Cogn 15:179–210
Vannucci M, Viggiano MP (2000) Category effects on the processing of plane-rotated objects. Perception 29:287–302
Vuilleumier P, Henson RN, Driver J, Dolan RJ (2002) Multiple levels of visual object constancy revealed by event-related fMRI of repetition priming. Nat Neurosci 5:491–499
Warrington EK, James M (1988) Visual apperceptive agnosia – a clinico-anatomical study of 3 cases. Cortex 24:13–32
Warrington EK, Taylor AM (1978) Two categorical stages of object recognition. Perception 7:584–694
Zago L, Fenske MJ, Aminoff E, Bar M (2005) The rise and fall of priming: how visual exposure shapes cortical representations of objects. Cereb Cortex 15:1655–1665

10 Interactions Between Shape Perception and Egocentric Localization

Hiroyuki Sogo[1,2] and Naoyuki Osaka[1]

1 Introduction

Neuropsychological studies of patients with damage to either the temporal or parietal region have suggested that these areas can be broadly divided into two functionally different pathways, a ventral "what" pathway for feature-related object vision and a dorsal "where" pathway for motor-oriented spatial vision (Milner and Goodale 1995; Mishkin and Ungerleider 1982). This is a reasonable separation since humans must resolve what an object is regardless of where it is to achieve object recognition, and vice versa to plan body actions in relation to the object. Neuroanatomical studies in the monkey cerebral cortex have demonstrated that parietal and temporal cortical areas are heavily connected with each other (Felleman and Van Essen 1991). Given the similarity between human and monkey cortical architecture (Van Essen 2003), it is expected that human temporal and parietal areas also have similar inter-connections. Such inter-connections would imply potential interactions between the temporal and parietal areas. However, it remains unclear how deeply these areas actually interact with each other. Concerning this question, we report recent studies suggesting that illusory perception of an object location called "saccadic compression of visual space" affects the perception of object shapes.

[1]Department of Psychology, Graduate school of Letters, Kyoto University, Yoshida-Honmachi, Sakyo, Kyoto 606-8501, Japan
[2]Institute for Human Science and Bio-medical Engineering, National Institute of Advanced Industrial Science and Technology, AIST Tsukuba Central 6, 1-1-1 Higashi, Tsukuba, Ibaraki 305-8566, Japan

2 Saccadic Compression of Visual Space

2.1 Localization of a Flash near the Time of Saccade Execution

In daily life, humans frequently make a voluntary rapid eye movement called a saccade. Although the retinal location of an object changes quickly when doing a saccade, we usually do not perceive this change as a movement of the object. A widely accepted explanation for this fact is that the object location that we perceive is represented with respect to the head (Bridgeman et al. 1994). It is necessary to calculate the head-centered representation to integrate retinal location of the object image to eye position. Denoting head-center representation, retinal image location and eye position as T, R and E, respectively, this calculation can be represented as T = R + E (Fig. 1a). Neurophysiological and neural network simulation studies have suggested that this calculation is performed in parietal cortical areas (Andersen and Zipser 1987). For example, strength of visual responses of neurons in Brodmann area 7a of monkey brain is modulated by eye position (Andersen et al. 1985, see Fig. 1b). Such property is typically found in the intermediate layer of a feed-forward neural network that calculates head-centered representation from retinal image and eye position signal (Zipser and Andersen 1988). Generally, representation of object location with respect to a body part of the observer such as head-center representation is called "egocentric representation". Egocentric representation of an object location is important for control of visually guided body actions, and integration of visual input with idiothetic or self-motion information (e.g., vestibular, motor efference copy and proprioception) is a common prerequisite for all kinds of egocentric representation.

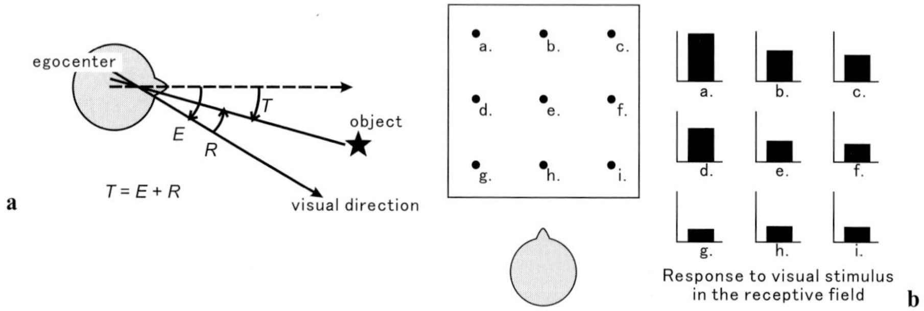

FIG. 1. Head-center representation of object location. Left: Head-centered representation of a visual object (*T*). T is the invariant of change in eye position when the object and head are fixed. Right: Eye-position modulation of visual responses of neurons in area 7a of the monkey brain. Visual responses during fixating on a ~ i are schematically shown in bar plots. Response of this model cell is enhanced when the monkey fixates on the top left. Head-center representation can be built up from this sort of eye-position-modulated visual response

Although head-center localization works well in daily life, it is known that the apparent location of a stimulus flashed within tens of milliseconds before, during or after saccade onset changes drastically depending on stimulus onset time relative to saccade onset time (Honda 1989, 1990, 1991; Matin et al. 1969, 1970). In general, localization error is in the same direction as the saccade when the flash is presented before saccade onset, and opposite to the saccade when the flash is presented after saccade. These results are interpreted as indicating that eye position signal changes more slowly compared to the actual change of eye position during saccade execution (Honda 1990). If this interpretation is correct, we could assume that localization error will be independent of the physical properties of a flash. However, previous studies showed that the presence of a luminous background causes dependency of the mislocalization size on retinal location of the flash (Bischof and Kramer 1968; Honda 1995; O'Regan 1984; Ross et al. 1997). Among these studies, Ross et al. (1997) reported that mislocalization strongly depended on the location of a flash when a green stimulus was flashed on an equiluminant red background. Figure 2 schematically shows the procedure and the results of their experiment. The subject made a horizontal saccade from F to T, while a vertical bar was presented for 10 ms before, during or after the saccade onset. Possible bar locations were –10 deg, 0 deg or 10 deg. The subject was asked

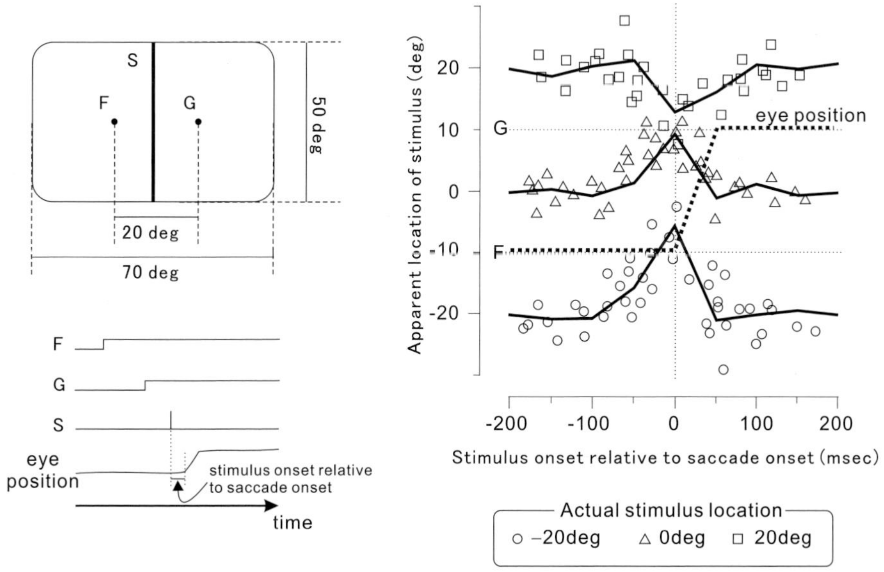

FIG. 2. Saccadic compression of visual space. Left: Spatial configuration of the stimulus and the time course of stimulus presentation. The subject made a saccade from F to G and reported the apparent location of the vertical bar (*S*). Right: Apparent position of vertical bars, plotted against stimulus onset time relative to saccade onset time. Each symbol *(circle, triangle or square)* corresponds to the result of a single trial

to report verbally where the vertical bar was perceived after saccade termination by indicating a horizontal ruler presented on the screen. The right panel of Figure 2 plots the apparent location of the vertical bar against the stimulus onset time relative to saccade onset. Each symbol corresponds to a single trial. This plot shows that the vertical bar was mislocalized as if visual space was compressed toward the goal of the saccade near the time of saccade onset (Ross et al. 1997). Ross et al. called this phenomenon "saccadic compression of visual space". Hereafter, we will abbreviate this "saccadic compression".

To date, it has been shown that saccadic compression also occurs in a luminance-modulated stimulus (Morrone et al. 1997). The effect of saccadic compression becomes stronger when stimulus contrast decreases (Michels and Lappe 2004). Compression in the direction orthogonal to the saccade is also observed, but the amount of compression is much smaller in comparison to that in the direction parallel to the saccade (Kaiser and Lappe 2004). Lappe et al. (2000) examined the dependence of perisaccadic mislocalization on the availability of visual spatial references at various times around a saccade. Their results showed that presaccadic compression occurs only if visual references are available immediately after, rather than before or during, the saccade. This finding indicates the importance of the time course of visual input on the generation of saccadic compression, while it is known that rapid displacement of a visual frame of reference simulating saccadic eye movement does not produce localization error similar to saccadic compression (Honda 1995; Morrone et al. 1997). These results suggest that saccade execution also plays an essential role in generating the compression effect.

2.2 *Neural Correlates of Saccadic Compression*

Krekelberg et al. (2003) suggested that the middle temporal (MT) and medial superior temporal (MST) areas may be concerned with the generation of saccadic compression. Initially, they measured the conditional probability of a particular firing rate for MT, MST, LIP and ventral intraparietal (VIP) neurons given the presentation of a flash at a particular location during fixation (this conditional probability was called "codebook"). The flash was presented during fixation or within ±200 ms from saccade onset, with "fixation codebook" obtained from data in the former condition and "perisaccadic codebook" obtained from the latter. They then examined how precisely the flash location can be estimated by translating the firing rate of neurons into stimulus location using these codebooks. The results indicate that MT and MST neurons can reliably encode retinal location of the flash with the fixation codebook. Performance of the perisaccadic codebook was no better than that of the fixation codebook even for decoding the location of perisaccadic flash. Most importantly, retinal location of the flash estimated using the fixation codebook was widely mislocalized in a manner similar to saccadic compression. From these findings, Krekelberg et al. suggested that dorsal downstream areas relying on MT and MST for retinal location information would inherit this mislocalization.

3 Saccadic Compression and Shape Perception

3.1 *Effects of Saccadic Compression on Shape Perception*

Ross et al. (1997) performed several experiments to argue that the illusion that they found results from compression of the neural representation of visual space. In one of these experiments, they presented some photographs of natural scenes 25 ~ 0ms before saccade execution, and asked the subjects to report verbally how the shape of objects in the scene were perceived. The result was that 11 of 13 subjects reported shape distortion of objects. Santoro et al. (2002) examined the effect of saccadic compression on the detection of a Glass pattern, i.e., a moiré pattern constructed from spatially random dots by duplication and displacement (Glass 1969). The subject made a horizontal saccade of 19deg amplitude and a stimulus was presented 25 ~ 0ms before saccade onset. The upper or bottom half of the stimulus was a horizontal or vertical Glass pattern, and the other half was random dots. Duration of the stimulus presentation was 5ms. The subjects reported which of the upper or bottom half was a Glass pattern. In the control condition, the subject judged the same stimulus without making a saccade. The results showed that detection of the horizontal Glass pattern was improved when it was presented before saccade onset, while there was no such improvement for the vertical Glass pattern (Fig. 3). Santoro et al. (2002) discussed that saccadic compression apparently shortened the horizontal dot separation resulting in

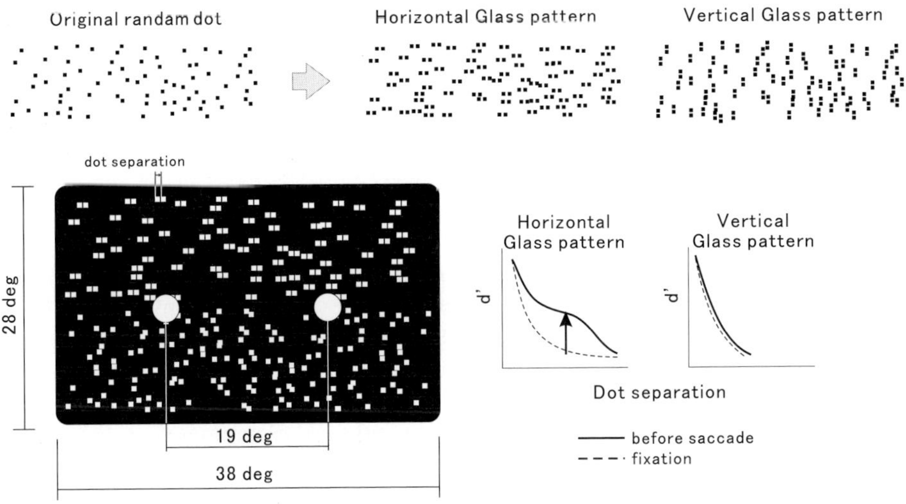

FIG. 3. Saccadic compression improves detection of Glass patterns. The subject made a horizontal saccade of 19deg amplitude and a dot pattern was flashed for 5ms. The subject then judged whether the top or bottom half of the dot pattern was a Glass pattern. They reported that performance of detection of the Glass pattern (*d'*) was improved when a horizontal Glass pattern was flashed immediately before saccade onset

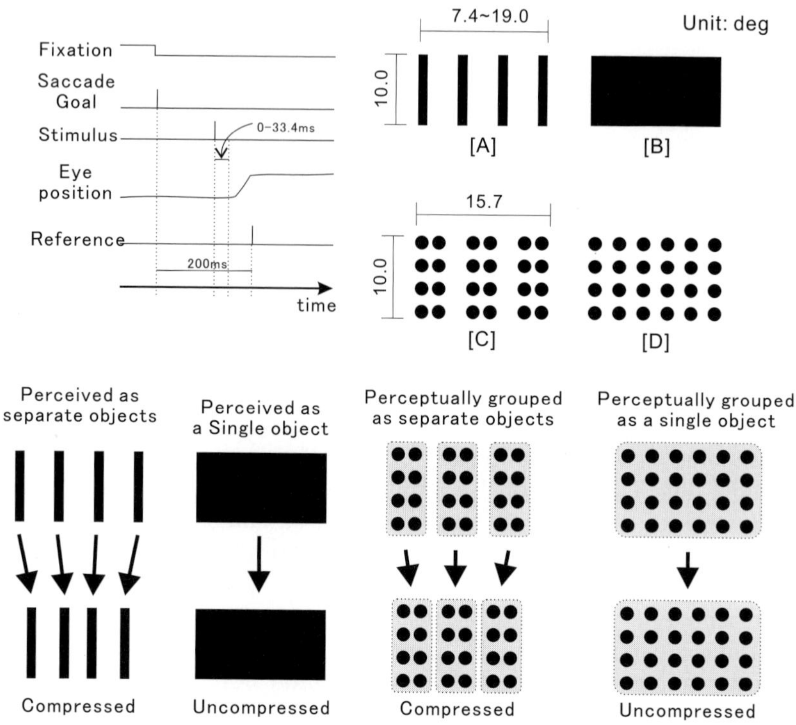

Fig. 4. Saccadic compression did not affect the apparent width of rectangle and proximity-grouped object. Top: The stimuli used in the experiment and the time course of stimulus presentation. Bottom: summary of the results and explanations by Matsumiya and Uchikawa (2001). A perceptually-grouped single object is uncompressed in the same manner as a solid object

improved detection of the horizontal Glass pattern. Considering the involvement of MT/MST in saccadic compression (Lappe et al. 2000), these findings suggest that transient changes in neural responses in MT/MST affect perception of object shape and global patterns as well as perception of object locations.

Contradictory to this suggestion, Matsumiya and Uchikawa (2001) reported that the apparent width of a rectangle presented before saccade onset was not compressed. Figure 4 shows their stimuli and procedure. At first, they compared the apparent widths of multiple bars (Fig. 4a) and solid rectangles (Fig. 4b) briefly presented before saccade onset. The subjects made a horizontal saccade to the location at which the saccade target was presented (20 deg right from the fixation point), and the stimulus was presented for one video frame (15.0 ~ 16.7 ms) so that the stimulus onset relative to saccade onset was 33.4 ~ 0 ms. The center of the stimulus was on the goal of the saccade. Two hundred milliseconds after the saccade goal was extinguished, a reference triangle was presented for one video frame. The subject judged whether the stimulus was smaller or larger than the reference. In control trials, the subject observed the same sequence of visual stimuli without making a saccade. The results showed that the apparent width of multiple bars was compressed compared to that in control trials, while the apparent width of a rectangle remained unchanged. In addition, they examined the apparent width of figures shown in Figure 4c,d. The procedure was the same as described above except that the width of the stimulus was fixed and the width of the reference rectangle was changed between trials. The results were that the apparent width of Figure 4c was compressed while that of Figure 4d remained unchanged. Based on these results, Matsumiya and Uchikawa (2001) suggested that shape perception of a single object is not distorted by saccadic compression. The "single object" need not be a solid, but a global pattern of multiple elements that is perceptually grouped as a single object (see the bottom part of Fig. 4) also defends against saccadic compression. This suggestion conflicts with the finding of saccadic compression of objects in a natural scene reported by Ross et al. (1997). Concerning this point, they speculated that there was no distortion of the object images in natural scenes, but the apparent location of each object image in the natural scenes shifted toward the saccade goal just before the saccades. The impression that the natural scene had become deformed would result from an apparent shift of each object image (Matsumiya and Uchikawa 2001).

3.2 Does Kanizsa Figure Defend against Saccadic Compression?

The suggestion by Matsumiya and Uchikawa (2001) further implies that perception of an object shape may be protected from transient changes in neural responses in MT/MST. Considering the theory of two visual pathways for "what" and "where" vision, this is an attractive hypothesis. However, Matsumiya and Uchikawa (2001) only showed that a solid object and a set of multiple objects organized by the so-called "Gestalt law of proximity" are uncompressed. To demonstrate that shape perception is truly protected from saccadic compression, it is necessary to show that other shape perception processes are also unaffected by saccadic compression. To investigate this point, Sogo and Osaka (2005) examined whether a Kanizsa-type subjective figure is protected against saccadic compression. The top left of Figure 5 shows the stimuli used in our experiment. "Disks" and "Pacmen" were expected to be apparently compressed. "Real Contour" and "Filled" were expected to remain uncompressed because these figures contained a single wide rectangle. Our question is whether the "Illusory contour" of a rectangle defined by a Kanizsa-type subjective contour (Kanizsa 1979) would be compressed or not.

The top right of Figure 5 shows the spatial configuration of the stimuli used in our experiment. The subject fixated on F at the beginning of a trial and a cross (G) was flashed for 20 ms 20 deg right to the F. The subject made a horizontal saccade as quickly as possible to the location where G was flashed. At a random time within 120 ~ 240 ms from the onset of G, one of the target stimuli shown in the top left of Figure 5 was presented for 10 ms. Width of the target was 16 deg

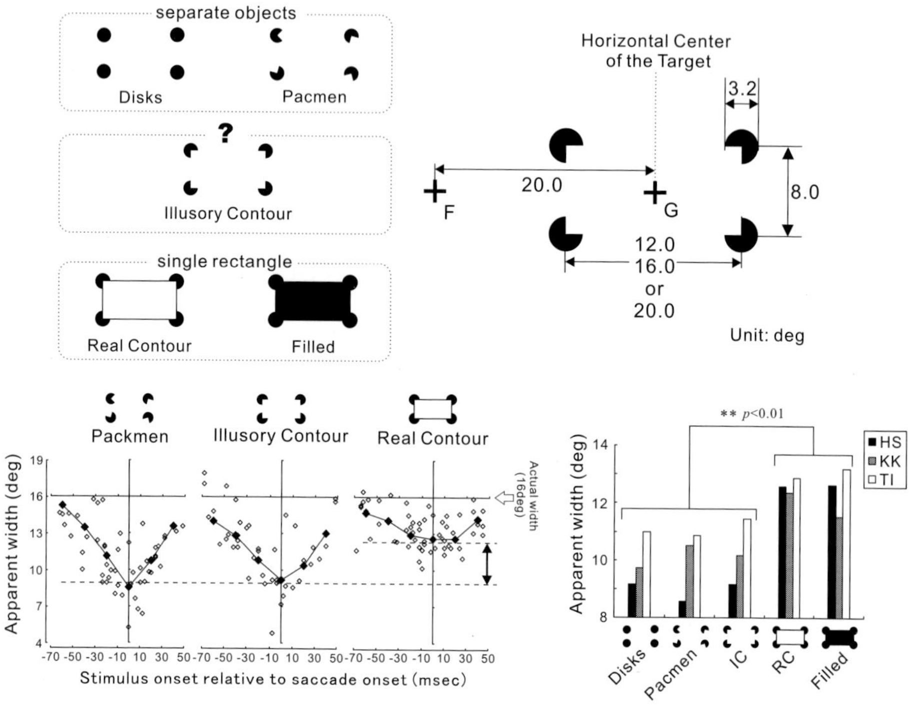

FIG. 5. Effect of saccadic compression on the apparent width of a Kanizsa figure. Top: The stimuli used in the experiment and spatial configuration of the stimuli. Bottom left: Representative results from a single subject. Bottom right: The minimal apparent width (average of the results of four subjects). The result for "Illusory Contour" was clearly similar to those for "Disks" and "Pacmen"

for 83% of the trials and 12 deg and 20 deg for the rest of the trials. A probe appeared approximately 1500 ms after onset of the saccade goal. The probe was the same shape and height as the target but the width of the probe was either wider or narrower than the target. The subject reported the apparent width of the target by adjusting the width of the probe using a joystick. Representative data are shown at the bottom left of Figure 5. The apparent width of all stimuli was narrowest when they were presented near the time of saccade onset. As expected, the minimum apparent width of "Pacmen" was narrower than that of "Real Contour". The result for "Illusory Contour" was clearly similar to that of "Pacmen". The bottom right of Figure 5 compares the minimal apparent widths of five targets (the mean from four subjects). The minimal apparent widths of "Disks", "Pacmen" and "Illusory Contour" were significantly narrower than those of "Real Contour" and "Filled". This finding indicates that Kanizsa-type illusory contour does not protect against saccadic compression. However, it is also possible to speculate that the horizontal distance between inducers (i.e., pacmen in the "Illusory Contour") might be too long to provoke a strong impression of illusory rectangle. To investigate this possibility, we examined whether the apparent width of a Kanizsa rectangle was compressed even when the horizontal distances between inducers were much shorter. The result was, interestingly enough, that minimal apparent width of "Illusory Contour" was not different from that of "Disks" for all distances examined (Sogo and Osaka 2005). Thus, we did not find any sign that perception of a Kanizsa figure is protected against saccadic compression.

3.3 Does a Line-Drawing of a Triangle Defend against Saccadic Compression?

An unexpected finding in the experiment shown in Figure 5 is that the apparent width of the "Real Contour" was also slightly compressed. This may be because this stimulus was not a single object in the strict sense but a compound of a rectangle and four disks. If the rectangle and four disks were perceived as separate objects and the apparent horizontal distance between disks was compressed, the overall width of the stimulus would be slightly compressed while the width of the rectangle was correctly perceived. Another possibility is that compression of the rectangle width was too small to detect with the method used by Matsumiya and Uchikawa (2001). To investigate this possibility, we examined the effect of saccadic compression on shape perception of a single object in a manner different from asking the subject to indicate the apparent width of the object (Sogo and Osaka 2007). Top left of Figure 6 shows the stimulus. The experiment consists of two conditions, "triangle" and "bar" condition. In the triangle condition, the subject fixated on F at the beginning of the trial and G was flashed for 20 ms. As quickly as possible, the subject made a horizontal saccade to the location where G had flashed. A triangle was flashed for 10 ms near the time of saccade onset. The triangle was randomly upright or upside-down, and the top or bottom vertex was offset from the horizontal center of the triangle. The task of the subject was to judge whether the top or bottom vertex was shifted to the left or right of the horizontal center of the rectangle. Under the "bar" condition, a vertical bar was flashed for 10 ms instead of the triangle. The location of the bar was randomly selected from three possible locations, indicated by B_L, B_R and B_C in Figure 6. The task for the "bar" condition was to point to the apparent location of the bar using a cursor. To compare the results of the "triangle" and "bar" condition, we calculated the distortion of the triangle under the "triangle" condition and the predicted distortion from mislocalization of the vertical bars (bottom left of Fig. 6). Distortion of the triangle in the "triangle" condition was defined as a proportion of the shift of the top or bottom vertex of the subjectively regular triangle (defined by 50% point of the psychometric function) from the horizontal center in proportion to the width of the triangle. The predicted distortion was defined as a shift of B_C location from the center of B_L, and B_R in proportion to the distance between B_L, and B_R. The right panel of Figure 6 shows the results for four subjects. Solid lines with filled squares show the observed distortion in the "triangle" condition, and dashed lines with open diamonds show the distortion

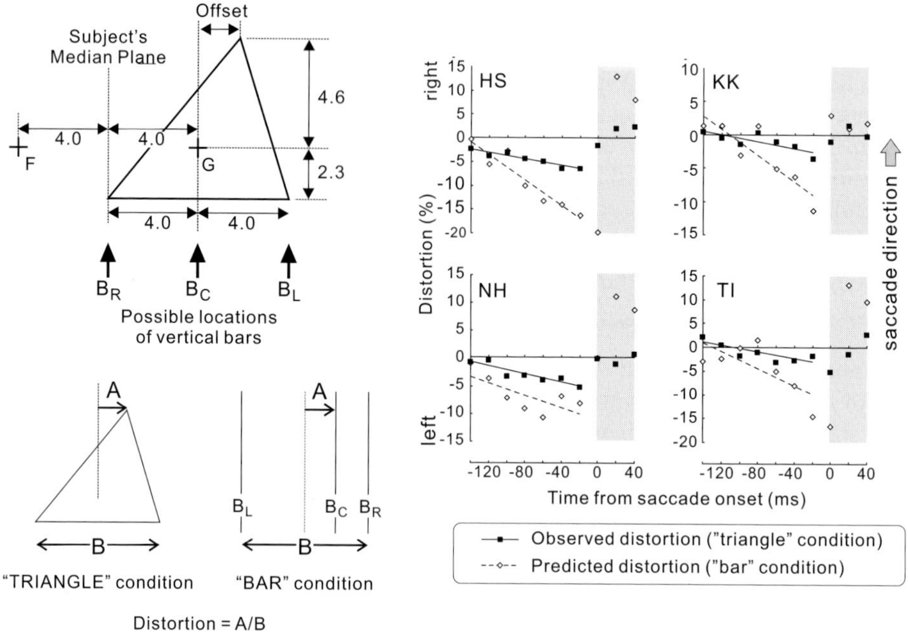

FIG. 6. Effect of saccadic compression on the perception of triangle shape. Top left: Spatial configuration of the stimuli. Bottom left: Definition of distortion. Dashed vertical lines indicates horizontal center of the stimuli. Right: The results from four subjects. Shadowed areas indicate that stimulus presentation and eye movement overlapped. These data should be unreliable because retinal image of the triangle was smeared in these trials

predicted from the "bar" condition. In the presaccadic period, both the observed and predicted distortion occurred in the direction opposite to the saccade and increased as the stimulus onset time relative to saccade onset became closer. The amount of the observed distortion was constantly smaller than that of the predicted distortion. These results suggest that the shape perception of a single object was less affected by saccadic compression compared to localization of vertical bars. In this sense, shape perception is protected from saccadic compression. However, this protection is not sufficient to eliminate all distortions of object shape.

3.4 How does Saccadic Compression Distort Shape Perception?

We have reviewed recent studies of saccadic compression and its effect on shape perception. These studies indicate that some figures are hardly affected by saccadic compression (simple geographic object and aproximity-grouped objects) while others are affected (Kanizsa figure and Glass pattern). Our tentative explanation for these findings is based on the following data and assumptions. Firstly, saccadic compression probably originates in transient changes of neural responses in the parietal areas (Krekelberg et al. 2003, for MT/MST neurons) and propagates to early visual processing areas through feedback connections (Deco and Lee 2004; Juan and Walsh 2003). Secondly, recognition of Kanizsa figures, Glass patterns and natural objects are processed in recurrent loop between early and higher visual areas (Grill-Spector et al. 2001; Kourtzi and Kanwisher 2001; Larsson et al. 1999; Mendola et al. 1999; Murray et al. 2002) and are somehow time-consuming (Brandeis and Lehmann 1989; Guttman and Kellman 2004; Murray et al. 2002; Ringach and Shapley 1996). Finally, proximity-based perceptual grouping is rapidly processed in early visual areas (Han et al. 1999, 2001). From these assumptions, we speculate that neural representations of single objects and proximity-grouped objects will be built up so quickly that these representations are hardly affected by feedback input from parietal areas where saccadic compression is generated. Compared to these, Kanizsa figures, Glass patterns and natural objects will be more strongly affected by feedback inputs from parietal areas because it takes a longer time to recognize these patterns and objects.

There may be other possible explanations for differences in the strength of compression effect between figures. For example, Sogo and Osaka (2005) pointed out that differences between representations of real and illusory contour in V1 and V2 (Ramsden et al. 2001) may cause stronger compression of a Kanizsa rectangle compared to that of a real rectangle. However, we consider it difficult to explain the effect of saccadic compression on shape perception without assuming that an interaction with the dorsal "where" pathway could have an effect on the ventral "what" pathway.

4 Conclusion

In this chapter, we showed new evidence supporting the interaction between dorsal and ventral pathways by showing that saccadic compression affects the shape perception process in the ventral pathway. The functional significance of such interaction is not clear at present. We speculate that this interaction may support building and maintaining representations of object shape under dynamic change of retinal images due to body actions, although a possible model showing how dorsal-ventral interaction achieves stable object representation could be advanced in the future.

References

Andersen RA, Essick GK, Siegel RM (1985) Encoding of spatial location by posterior parietal neurons. Science 230:456–458

Andersen RA, Zipser D (1987) The role of the posterior parietal cortex in coordinate transformations for visual-motor integration. Can J Physiol Pharmacol 66:488–501

Bischof N, Kramer E (1968) Untersuchungen und überlegungen zur richtungwahrnehmung bei willkürlichen sakkadischen augenbewegungen. Psychologische Forschung 32:185–218

Brandeis D, Lehmann D (1989) Segments of event-related potential map series reveal landscape changes with visual attention and subjective contours. Electroencephalogr Clin Neurophysiol 73:507–519

Bridgeman B, Van der Heijden AH, Velichkovsky BM (1994) A theory of visual stability across saccadic eye movements. Behav Brain Sci 17:247–292

Deco G, Lee TS (2004) The role of early visual cortex in visual integration: a neural model of recurrent interaction. Eur J Neurosci 20:1089–1100

Felleman DJ, Van Essen DC (1991) Distributed hierarchical processing in the primate cerebral cortex. Cereb Cortex 1:1–47

Glass L (1969) Moiré effects frfom random dots. Nature 243:578–580

Grill-Spector K, Kourtzi Z, Kanwisher N (2001) The lateral occipital complex and its role in object recognition. Vision Res 41:1409–1422

Guttman SE, Kellman PJ (2004) Contour interpolation revealed by a dot localization paradigm. Vision Res 44:1799–1815

Han S, Humphreys GW, Chen L (1999) Uniform connectedness and classical gestalt principles of perceptual grouping. Percept Psychophys 61:661–674

Han S, Song Y, Ding Y, Yund EW, Woods DL (2001) Neural substrates for visual perceptual grouping in humans. Psychophysiology 38:926–935

Honda H (1989) Perceptual localization of visual stimuli flashed during saccades. Percept Psychophys 45:162–174

Honda H (1990) Eye movements to a visual stimulus flashed before, during, or after a saccade. In: Jeannerod M (Ed) Attention and performance, Vol. XIII. Erlbaum, Hillsdale NJ, pp 567–582

Honda H (1991) The time courses of visual mislocalization and of extraretinal eye position signals at the time of vertical saccades. Vision Res 31:1915–1921

Honda H (1995) Visual mislocalization produced by a rapid image displacement on the retina: examination by means of dichoptic presentation of a target and its background scene. Vision Res 35:3021–3028

Juan CH, Walsh V (2003) Feedback to v1: a reverse hierarchy in vision. Exp Brain Res 150:259–263

Kaiser M, Lappe M (2004) Perisaccadic mislocalization orthogonal to saccade direction. Neuron 41:293–300

Kanizsa G (1979) Organization in vision. Springer, New York

Kourtzi Z, Kanwisher N (2001) Representation of perceived object shape by the human lateral occipital complex. Science 293:1506–1509

Krekelberg B, Kubischik M, Hoffmann KP, Bremmer F (2003) Neural correlates of visual localization and perisaccadic mislocalization. Neuron 37:537–545

Lappe M, Awater H, Krekelberg B (2000) Postsaccadic visual references generate presaccadic compression of space. Nature 403:892–894

Larsson J, Amunts K, Gulyas B, Malikovic A, Zilles K, Roland PE (1999) Neuronal correlates of real and illusory contour perception: functional anatomy with pet. Eur J Neurosci 11:4024–4036

Matin L, Matin E, Pearce DG (1969) Visual perception of direction when voluntary saccades occur. I. Relation of visual direction of a fixation target extinguished before a saccade to a flash presented during the saccade. Percept Psychophys 5:65–80

Matin L, Matin E, Pola J (1970) Visual perception of direction when voluntary saccades occur: ii. Relation of visual direction of a fixation target extinguished before a saccade to a subsequent test flash presented before the saccade. Percept Psychophys 8:9–14
Matsumiya K, Uchikawa K (2001) Apparent size of an object remains uncompressed during presaccadic compression of visual space. Vision Res 41:3039–3050
Mendola JD, Dale AM, Fischl B, Liu AK, Tootell RB (1999) The representation of illusory and real contours in human cortical visual areas revealed by functional magnetic resonance imaging. J Neurosci 19:8560–8572
Michels L, Lappe M (2004) Contrast dependency of saccadic compression and suppression. Vision Res 44:2327–2336
Milner AD, Goodale MA (1995) The visual brain in action, Vol. 27. Oxford University Press, Oxford
Mishkin M, Ungerleider LG (1982) Contribution of striate inputs to the visuospatial functions of parieto-preoccipital cortex in monkeys. Behav Brain Res 6:57–77
Morrone MC, Ross J, Burr DC (1997) Apparent position of visual targets during real and simulated saccadic eye movements. J Neurosci 17:7941–7953
Murray SO, Kersten D, Olshausen BA, Schrater P, Woods DL (2002) Shape perception reduces activity in human primary visual cortex. Proc Natl Acad Sci USA 99:15164–15169
O'Regan JK (1984) Retinal versus extraretinal influences in flash localization during saccadic eye movements in the presence of a visible background. Percept Psychophys 36:1–14
Ramsden BM, Hung CP, Roe AW (2001) Real and illusory contour processing in area v1 of the primate: a cortical balancing act. Cereb Cortex 11:648–665
Ringach DL, Shapley R (1996) Spatial and temporal properties of illusory contours and amodal boundary completion. Vision Res 36:3037–3050
Ross J, Morrone MC, Burr DC (1997) Compression of visual space before saccades. Nature 386:598–601
Santoro L, Burr D, Morrone MC (2002) Saccadic compression can improve detection of glass patterns. Vision Res 42:1361–1366
Sogo H, Osaka N (2005) Kanizsa figure does not defend against saccadic compression of visual space. Vision Res 45:301–309
Sogo H, Osata N (2007) Distortion of apparent shape of an object immediately before saccade. Spat Vis 20:265–276
Van Essen DC (2003) Organization of visual areas in macaque and human cerebral cortex. In: Chalupa LM, Werner JS (Eds) The visual neuroscience, Vol. 1. MIT Press, Cambridge, MA, pp 507–521
Zipser D, Andersen RA (1988) A back-propagation programmed network that simulates response properties of a subset of posterior parietal neurons. Nature 331:679–684

11
Feature Binding in Visual Working Memory

JUN SAIKI

1 Object Recognition and Visual Working Memory

Our visual world contains numerous objects. An important function of the visual system is to recognize an object by comparing perceptual and memory representations. Although we seldom have any problems in recognizing natural objects, which promotes the belief that object recognition is a quite simple process of matching perceptual and memory representations, recognition does in fact involve extremely complicated visual processing. The fact that objects are almost never presented in isolation illustrates the complexity and difficulty of object recognition. The cluttered nature of our visual environment poses an object segmentation problem (including figure/ground segregation problem), which itself is quite difficult. Even if one can successfully segment a set of objects, there is another problem for the visual system to solve: the so-called binding problem. If there are multiple objects, each of which has its own feature values such as shape, color, size, and so on, then how does the visual system properly maintain the correct correspondences of these features? This chapter focuses on this binding problem in both object recognition and visual working memory.

2 Binding in Object Recognition and Visual Working Memory

The binding problem emerges whenever a system needs to represent multiple entities having multiple features simultaneously. Thus, both the literature on recognition of a multi-part object and that on storage of multiple objects in short-term memory contain theoretical discussions on the binding problem. In the object recognition literature, Hummel and Biederman (1992) illustrated the importance of the binding problem by proposing a neural network model of

Graduate School of Human and Environmental Studies, Kyoto University, Yoshida-Nihonmatsucho, Sakyo-ku, Kyoto 606-8601, Japan

structural-description-based object recognition. In Biederman's (1987) RBC theory of object recognition, natural objects are represented by structural descriptions composed of a set of part representations and their spatial relations. Because parts and their relations are represented by a set of geometric properties and parts need to be simultaneously represented, it is crucial to properly represent the binding of features. Hummel and Biederman proposed a network using temporal synchrony of oscillatory activity to implement feature binding mechanism.

Somewhat in parallel, in visual working memory literature, where we appear to be able to maintain multiple objects simultaneously, a similar proposal has been presented by Luck and Vogel (1997). Luck and Vogel showed that a functional unit of visual working memory is object representations where their features are integrated, and that the capacity of visual working memory is about 3–5 objects. Based on these findings, Luck and Vogel proposed that synchronous neural oscillation binds visual features of an object (Todd and Marois 2004; Vogel et al. 2001; Vogel and Machizawa 2004; see also Cowan 2001 for a review). However, these previous studies are insufficient to provide strong evidence supporting the role of objects in visual memory. The results of some studies using a change detection task suggest that our capacity for object representation in visual memory is more limited than previously believed (Alvarez and Cavanagh 2004; Olson and Jiang 2002; Wheeler and Treisman 2002; Xu 2002). Moreover, Saiki (2002, 2003a, b) recently devised a paradigm called multiple object permanence tracking (MOPT) to investigate memory for binding in visual memory, and a reported similar limitation in visual memory.

3 Is Limited Binding Specific to Visual Working Memory?

Although the literature is currently equivocal regarding the capacity of memory for feature binding, it is likely that our memory for binding is not as powerful as Luck and Vogel first proposed. This raises a question of whether bindings in visual working memory and object recognition are fundamentally different, because given that we can easily recognize objects with 3–4 parts (Biederman 1987), the binding mechanism for object recognition appears to be able to deal with 3–4 parts simultaneously.

There are some important differences between the issues of binding in object recognition and visual working memory. Among these, I focus on one particular aspect in this chapter. Binding in object recognition is structural in the sense that a particular combination of component features is associated with a higher level description of parts, whereas binding discussed in visual working memory lacks such a higher level unit. For example, a combination of "straight axis, curved cross section, and constant size of cross section" defines a geon of cylinder, whereas a combination of "red, square, and large" does not have any label for it.

In other words, a problem with stimuli used in visual working memory may be the lack of such structural relations. To address this issue, an experiment was conducted to compare the effect of higher level nodes on maintenance of feature binding in visual working memory. Two specific questions are addressed:

(1) Does pre-stored knowledge about shape-color correspondence facilitate memory for feature bindings, and if so, how?
(2) Does constant mappings of shape-color correspondence within an experimental session facilitate memory for feature bindings, and if so, how?

If manipulations of (1) or (2) facilitate performance, the limited capacity for feature bindings in previous works is likely to reflect the arbitrary and independent nature of feature conjunctions used in the experiments. In contrast, if the factors above do not facilitate performance, then the capacity limit is likely to be more general.

4 MOPT as a Paradigm to Investigate Binding in Visual Working Memory

Saiki (2002, 2003a, b) recently devised a paradigm called multiple object permanence tracking (MOPT) to investigate whether humans can track multiple object files in a dynamic situation, and showed that object motion, even if slow and easily tracked, severely disrupts the ability to maintain multiple object files. In the MOPT task, four to six objects with different colors or shapes are placed at equal eccentricity, then rotated behind a windmill-shaped occluder (Fig. 1). In the middle of the rotation sequence, features of two objects may be switched during an occlusion. The task of the observer is to detect whether a feature switch occurred. The speed of disk rotation was manipulated by the relative motion of disks and occluder, to investigate the effect of motion in a parametric manner. In general, switch detection was markedly impaired as motion speed increased (Saiki 2002, 2003a, b).

Two necessary conditions must be met to properly evaluate the use of feature conjunctions. First, to eliminate possible contributions from simple feature information, the stimulus set should use identical sets of features in different combinations. The second condition is the use of a task being able to evaluate the representation of feature combination. MOPT paradigm in Saiki (2003a, b) only satisfied the first condition. Change detection tasks used in visual memory, including the original MOPT, fail to satisfy the second condition, because simple stimulus salience can account for correct change detection without using memory for feature combinations. One task satisfying the second condition is the perceptual identification task used in perceptual feature binding (Ashby et al. 1996). However, the simple identification task also contains problems. Even if subjects are simply asked to report an object with change, the cognitive load in response

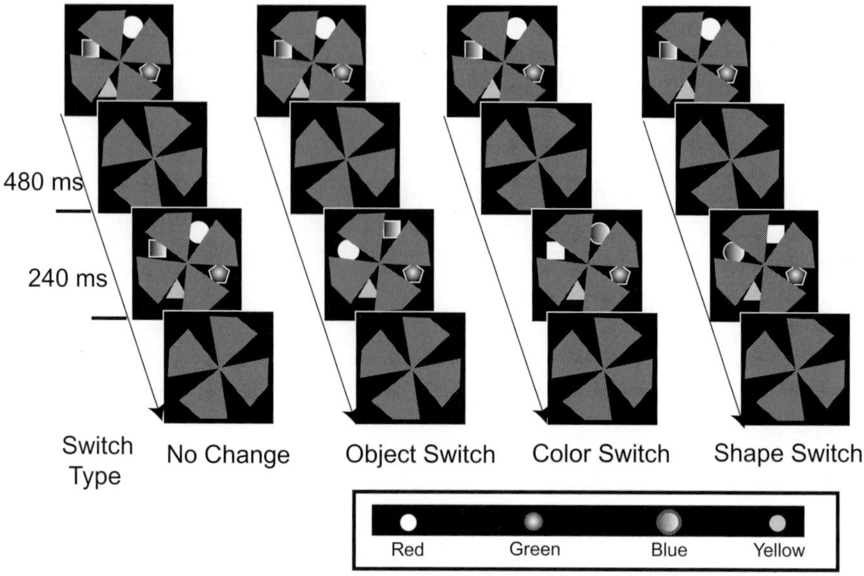

FIG. 1. Schematic illustration of the multidimensional multiple-object permanence tracking (MOPT) task. In this example, the objects are stationary and the occluder is rotating. In the second visible period, four types of switches may occur between the circle and the square

mapping is significant, and is quite likely to affect memory performance. Furthermore, direct identification forces participants to transform visual information into a verbal form, which can also compromise visual memory performance.

To avoid problems with both detection and simple identification tasks, a task called type identification was devised (Saiki and Miyatsuji, 2007). The paradigm can be illustrated using a color and shape example. Suppose the first display contains a red square on the left and a blue circle on the right, then four possible change types exist: no change (red square and blue circle); color change (blue square and red circle); shape change (red circle and blue square) and object change (blue circle and red square). The type identification task requires participants to identify which event occurs in the stimulus sequence, as discrimination among four alternatives. Correct identification of change type requires memory for feature combinations. At the same time, unlike simple identification, the cost in response mapping is negligible. These characteristics are crucial, particularly when using the wide varieties of colors and shapes seen in most visual memory tasks. If several colors and shapes are used, type identification based solely on salience is almost impossible, and the cost in terms of response mapping in simple identification becomes prohibitive. Compared with change detection tasks, the type identification task can thus extract important additional information regarding binding memory.

To use the type identification paradigm, multiple object features such as color and shape must be used in the context for which the spatiotemporal locations of the objects are relevant. Otherwise, change detection and type identification become identical. In this study, objects were defined by conjunction of color and shape.

Using the multidimensional MOPT paradigm with the type identification procedure, the roles of prestored memory representations of color-shape conjunctions in maintaining object information in visual working memory were evaluated. An experiment was conducted to investigate (1) whether known color-shape conjunctions facilitate maintenance of multiple object representations in visual working memory, (2) whether fixed color-shape conjunction facilitates maintenance of multiple object representations, and (3) whether patterns of errors demonstrate the roles of prestored conjunctions in visual working memory.

5 Experiment

5.1 Method

5.1.1 Participants

The experiment used 12 participants, and all displayed normal color vision.

5.1.2 Design

Two main independent variables were object type and motion type. The object types were natural (N) when natural objects were used, geometric-constant (GC) when geometric figures were used as in previous studies, while the shape-color correspondences were fixed, and geometric-varied (GV), which is identical to previous studies (Fig. 2). The motion types were object motion and occluder motion.

5.1.3 Materials

Participants were shown a pattern of four colored objects and an occluder on top. Smooth rotation of the pattern and occluder at constant angular velocities resulted in alternating appearance and disappearance of the pattern. The four colored objects were configured in a diamond pattern, with each object placed at a visual angle of 4.0° from the center of the occluder. Shapes used for objects in the geometric conditions were circle, square, hexagon and triangle. Objects used in the natural condition were lobster, frog, banana, and violin, which had clear associated colors, based on a preliminary survey. Colors were those typically associated colors: red, green, yellow, and brown, for both natural and geometric conditions. The colored objects were occluded using a gray windmill-shaped occluder ($18.3\,cd/m^2$), and the background was light gray ($28.9\,cd/m^2$). The sequence was either regular clockwise or counterclockwise rotation throughout, containing one visible period in which the locations of features of the two objects

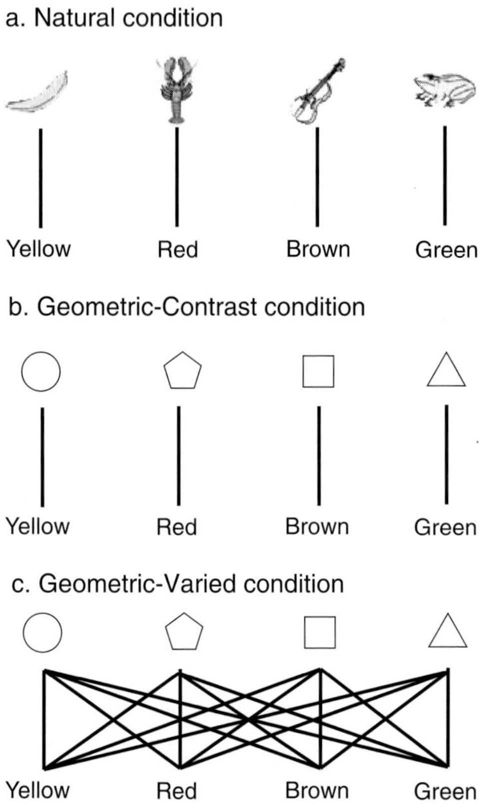

FIG. 2. Schematic illustration of the manipulation of object type. **a** Natural condition had fixed and natural correspondences between shape and color, **b** Geometric-constant condition had arbitrary, but constant correspondences between shape and color, **c** Geometric-varied conditions had arbitrary shape-color correspondences varying across trials

were switched. A total of four events were possible: object-switch with simultaneous switch of color and shape; color-switch alone; shape-switch alone; and no switch (Fig. 1). The occluder displayed four openings of 30°, through which the colored pattern could be seen. A switch event occurred between the 3rd and 6th occluded periods. When a switch event occurred, the sequence terminated at the next occlusion period. The timing of sequence termination under the no switch condition matched that under other conditions. Time and location of switches were unpredictable to the participants. Participants were asked to identify event types without feedback as to which was correct.

Object motion was manipulated by the relative motion of the pattern and occluder, as described by Saiki (2003b). In the occluder motion condition, objects were stationary and the occluder rotated at 126°/s. In the object motion condition, the object rotated at 84°/s, and the occluder rotated at 42°/s in the opposite direction. Note that both conditions had exactly the same duration of visible period (480 ms) and occluded period (240 ms). Experimental programs were written in MATLAB, using Psychophysics Toolbox extensions (Brainard 1997; Pelli 1997).

5.1.4 Procedure

Each experimental trial began with presentation of the sequence, followed by the appearance of four response boxes for event types. Participants selected responses by clicking a response box. To avoid verbal encoding of color and shape, articulatory suppression was achieved by getting participants to repeatedly say "da, da, da". The entire experiment comprised three experimental sessions, each containing 192 trials. Object type condition was fixed throughout each experimental session, and order of sessions was counterbalanced across participants. Within each session, object motion conditions were randomly mixed from trial to trial. In each object type session, each motion condition comprised 96 trials, with 24 trials for each event type, for a total of 576 experimental trials.

5.2 Results

5.2.1 Correct Type Identification

Figure 3 shows the proportions of correct type identification as a function of object type and object motion conditions. Analyses of proportions of correct data used arcsine transformed value as dependent variables. First, ANOVA with a 3

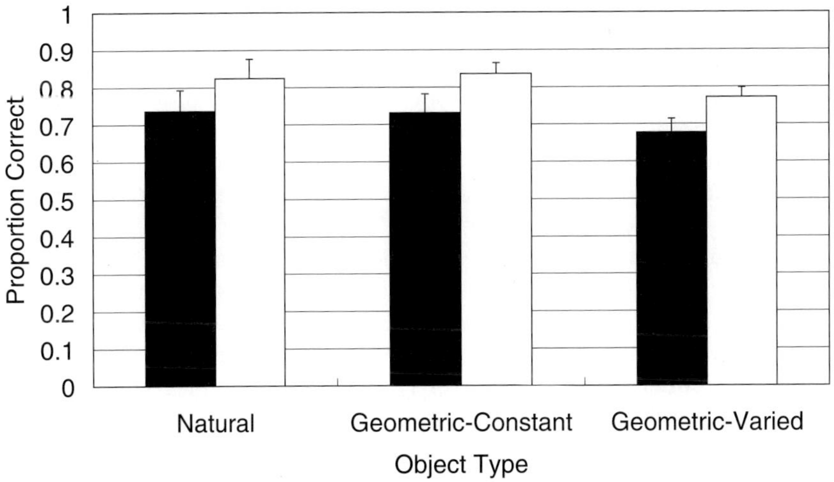

Fig. 3. Mean proportion of correct identifications for stationary and motion conditions as a function of object types; *black bars*: object motion, *white bars*: occluder motion. Error bars denote standard errors of the mean

(object type) × 2 (object motion) × 4 (switch type) design was conducted for the proportion of correct identification. Here I focused on results involving the object type. The main effects of object type ($F(1,4) = 11.43$, $P < 0.05$) were significant, although Schéffe's multiple comparison did not demonstrate significant pairwise differences. Differences among object type conditions were rather weak. The object type showed significant interaction with switch type ($F(3,12) = 24.53$, $P < 0.0001$), and three-way interaction was also significant ($F(3,12) = 4.89$, $P < 0.05$).

5.2.2 Response Type Analyses

Although object type did not show strong effects on the overall proportion correct, its significant interaction with other factors suggests that object type modulates the task performance significantly. To clarify these effects, I next analyzed patterns of errors. To show the effects of object type clearly, two types of events were defined: type variant event, and type invariant event. Type variant events composed of shape-switch and color-switch, where the combination of color and shape differs before and after the switch. Type invariant events composed of object-switch and no-switch, where the combination of color and shape does not change across the switch. First, I analyzed the type variant trials. Errors found in these trials were classified into the following three categories: feature miss, feature confusion, and feature false alarm. Feature misses and false alarms are errors where observers responded "no-switch" and "object-switch", respectively, because one feature-switch event is either missed or falsely reported. Feature confusions are errors between color-switch and shape-switch. Figure 4a shows mean proportions of these error types for object type conditions. It is clear that the natural condition showed significantly more feature confusion errors than the other conditions, ($F(4,44) = 16.52$, $P < 0.0001$). Furthermore, as shown in Figure 4b, not only the frequency but also the direction of confusion showed significant differences. The natural condition showed a strong asymmetry such that confusion of shape as color is much more frequent than the other direction, whereas the other two conditions did not show such a difference, ($F(2,22) = 30.42$, $P < 0.0001$).

Next, I analyzed type invariant trials. Errors in these trials were classified into two categories: Feature errors and location errors. Feature errors were selections of color- or shape-switch events for type invariant events, and location errors were confusion between object-switch and no-switch. Figure 5 shows mean frequencies of these error types for object type conditions. Unlike the type variant trials, the type invariant trials showed a significant difference according to whether color-shape mapping remained constant or not. Namely, the geometric-varied condition showed significantly higher frequency of feature errors than the other two conditions, ($F(2,22) = 22.50$, $P < 0.0001$). Error patterns in the natural and geometric-constant conditions were quite similar to each other.

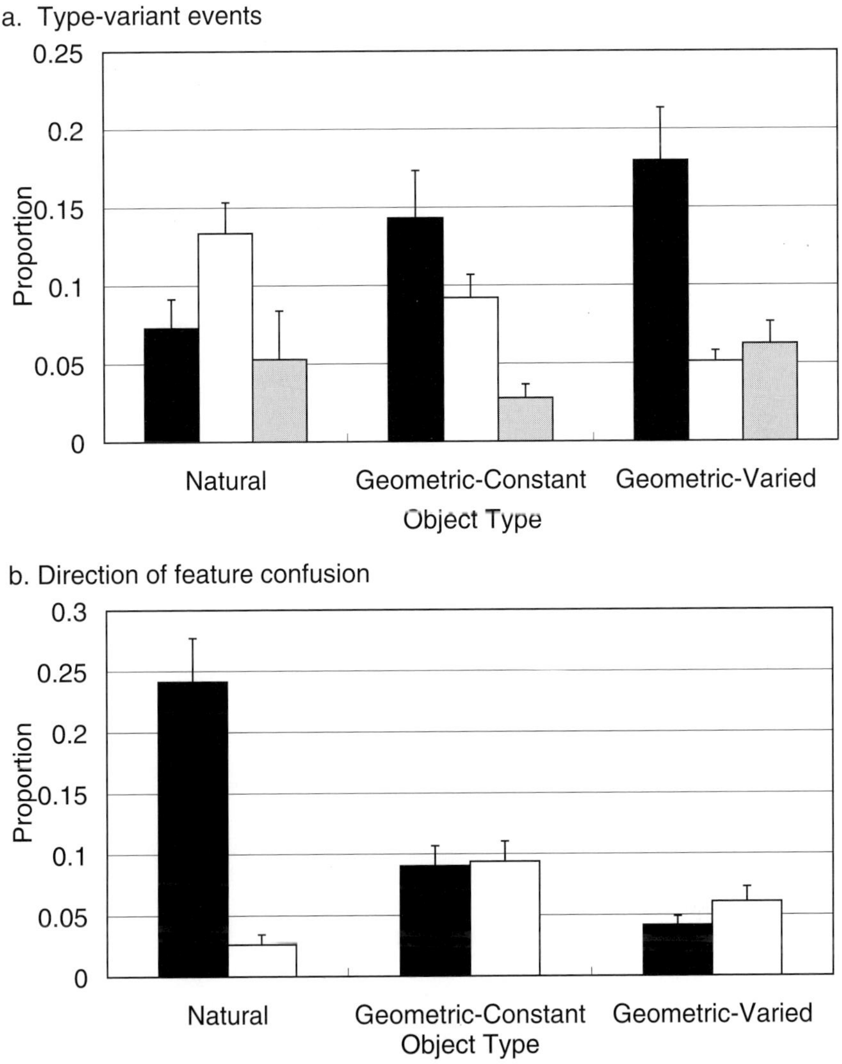

FIG. 4. **a** Mean proportion of errors for type variant trials as a function of object type conditions, *black bar*: feature miss, *white bar*: feature confusion, *gray bar*: feature false alarm, **b** Mean proportion of subtypes of feature confusion errors, *black bars*: shape switch judged as color switch, *white bars*: color switch judged as shape switch

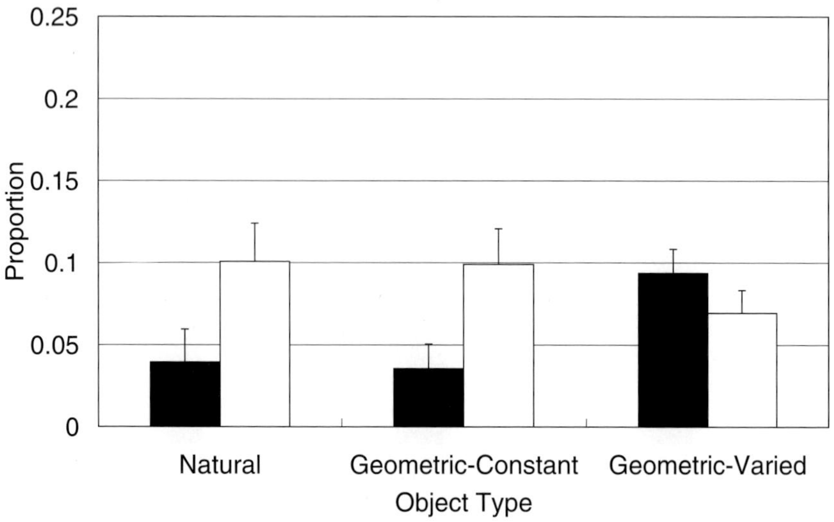

FIG. 5. Mean proportion of errors for type invariant trials as a function of object type conditions, *black bars*: feature error, *white bars*: location error

6 General Discussion

Multidimensional MOPT with the type identification paradigm demonstarated that memory for feature binding is severely limited, which is consistent with previous MOPT experiments and other studies. This chapter investigated whether this limitation is specific to the use of arbitrary combinations of color-shape, and the answer was no. Two additional conditions, the natural condition where prestored color-shape conjunction is available, and the geometric-constant condition, where color-shape correspondence is fixed throughout the experiment, showed only a weak tendency toward performance improvement, and these conditions showed severe performance impairment under the moving condition. The natural object and geometric-constant conditions were virtually the same in accuracy, suggesting that prestored color-shape conjunctions had limited effect on percent correct data.

However, analyses of error types demonstrated strong effects of prestored conjunction on task performance. Compared with geometric conditions, the natural condition showed significantly more errors confusing between color-switch and shape-switch, suggesting that observers were quite sensitive to detect a change in object identity, but not able to accurately identify the switch type. In the natural condition, color and shape form a unit of object identity, but to identify the switch type, its component (either color or shape) and location needs to be bound. Observers can detect the occurrence of color or shape switch when they see a green lobster, but they are not good at telling whether a red lobster changed to a green lobster (i.e., color switch), or a green frog changed to a green lobster (i.e., shape switch). In fact, they had a strong bias to judge any switch involving identity change as a color switch.

In contrast, although the error rates were about the same under the geometric-constant condition, the pattern of errors is quite different. As shown in Figure 4, color and shape behave more independently, even when the conjunctions are completely fixed. Unlike the case of lobster, when a predefined red-square combination changed to a red-circle (i.e., shape-switch), errors were more likely to be an indication of no switch (i.e., overlooking the shape-switch), and in the case of feature confusion, errors occurred evenly in both directions.

Results for the natural condition support a view that visual features are first bound together to form a type representation, before further binding to a spatiotemporal location to form a token (Kanwisher 1991). Moreover, this view holds only when type information is prestored in LTM, and without prestored types, shape-color conjunctions played no significant role. Types may be representations that are prestored in the long-term memory system, with arbitrary color-shape combinations not prestored in LTM as types.

More importantly, the availability of type information did not facilitate task performance in MOPT. This raises a possibility that even the feature binding in structural descriptions may have a similar limitation. As Hummel and Biederman (1992) described, structural description is not simply a co-activation of a set of geons, but also a binding of parts with relations. Given part representation is a set of its components, it is similar to the type representation discussed here. Thus, structural description needs binding of parts (types) with spatial information, which corresponds to the binding of types with their locations in MOPT task. Thus, the formal structure has a certain level of similarity between multiple objects in the MOPT task and an object's structural description. In addition, recent neuroimaging studies using MOPT (Imaruoka et al. 2005) and part-combined objects (Hayworth and Biederma, 2005) suggests spatial information processing in both cases using the same neural substrates, posterior parietal cortex.

However, there are important differences as well. For example, parts are tightly grouped by connectedness and other grouping factors (Saiki and Hummel 1998), but objects are completely separated in MOPT. Binding in structural description formation is limited to shape information, but shape and color (and other object features) are used in MOPT. Clearly, how these factors affect binding performance is an issue for further studies, but this chapter shows that limits in feature binding in visual working memory are not simply an artifact of arbitrary feature combinations, and these limits may have a broader common ground including binding in object recognition.

Acknowledgements. I wish to thank Toshio Inui for helpful comments and technical assistance, and Hirofumi Miyatsuji for data collection. This work was supported by Grants-in-Aid for Scientific Research for JMEXT (No. 13610084), The

Research for the Future Program from JSPS (JSPS-RFTF99P01401), the 21st Century COE Program from MEXT (D-2 to Kyoto University), and PRESTO "Intelligent Cooperation and Control" from Japan Science and Technology Agency (JST).

References

Alvarez GA, Cavanagh P (2004) The capacity of visual short-term memory is set both by visual information load and by number of objects. Psychol Sci 15:106–111

Ashby FG, Prinzmetal W, Ivry R, Maddox WT (1996) A formal theory of feature binding in object perception. Psychol Rev 103:165–192

Biederman I (1987) Recognition-by-components: a theory of human image understanding. Psychol Rev 94:115–147

Brainard DH (1997) The psychophysics toolbox. Spat Vis 10:443–446

Cowan N (2001) The magical number 4 in short-term memory: a reconsideration of mental storage capacity. Behav Brain Sci 24:87–185

Hayworth KJ, Biederman I (2005) Differential fMRI activity produced by variation in parts and relations during object perception. J Vis 5(8):740

Hummel JE, Biederman I (1992) Dynamic binding in a neural network for shape recognition. Psychol Rev 99:480–517

Imaruoka T, Saiki J, Miyauchi S (2005) Maintaining coherence of dynamic objects requires coordination of neural systems extended from anterior frontal to posterior parietal brain corteces. NeuroImage 26:277–284

Kanwisher NG (1991) Repetition blindness and illusory conjunction: errors in binding visual types with visual tokens. J Exp Psychol Hum Percept Perform 17:404–421.

Luck SJ, Vogel EK (1997) The capacity of visual working memory for features and conjunctions. Nature 390:279–281

Olson IR, Jiang Y (2002) Is visual short-term memory object based? Rejection of the "strong-object" hypothesis. Percept Psychophys 64:1055–1067

Pelli DG (1997) The video toolbox software for visual psychophysics: transforming numbers into movies. Spat Vis 10:437–442

Saiki J (2002) Multiple-object permanence tracking: limitation in maintenance and transformation of perceptual objects. In: Hyona J, Munoz DP, Heide W, Radach R (Eds) The Brain's eye: neurobiological and clinical aspects of oculomotor research (progress in brain research), Vol. 140. Elsevier Science, Amsterdam, pp 133–148

Saiki J (2003a) Feature binding in object-file representations of multiple moving items. J Vis 3:6–21

Saiki J (2003b) Spatiotemporal characteristics of dynamic feature binding in visual working memory. Vision Res 43:2107–2123

Saiki J, Miyatsuji H (2007) Feature binding in visual working memory evaluated by type identification paradigm. Cognition 102:49–83

Saiki J, Hummel JE (1998) Connectedness and the part-relation integration in shape perception. J Exp Psychol Hum Percept Perform 24:227–251

Todd JJ, Marois R (2004) Capacity limit of visual short-term memory in human posterior parietal cortex. Nature 428:751–754

Vogel EK, Machizawa MG (2004) Neural activity predicts individual differences in visual working memory capacity. Nature 428:748–751

Vogel EK, Woodman GF, Luck SJ (2001) Storage of features, conjunctions and objects in visual working memory. J Exp Psychol Hum Percept Perform 27:92–114

Wheeler ME, Treisman A (2002) Binding in short-term visual memory. J Exp Psychol Gen 131:48–64

Xu Y (2002) Limitations of object-based feature encoding in visual short-term memory. J Exp Psychol Hum Percept Perform 28:458–468

12 Biased Competition and Cooperation: A Mechanism of Mammalian Visual Recognition?

GUSTAVO DECO[1,2], MARTIN STETTER[3], and MIRUNA SZABO[3,4]

1 Introduction

In humans and mammals with higher cognitive capabilities, the neocortex is a very prominent brain structure (Fig. 1). As such it seems to be crucially involved in the cognitive processes. The neocortex can be subdivided into a set of functionally different areas (Van Essen et al. 1992), and it communicates with most of the other brain systems. It is a structure with a high internal functional complexity and diversity which is involved in most aspects of cerebral processing. Various cortical areas represent and process different aspects of the environment and the subject's internal states in a distributed way. In the visual modality for example, occipital to temporal regions of the brain are thought to mainly represent object identity-related sensory information, whereas occipital to parietal brain regions are thought to mainly represent and process spatial information and aspects preparing motor plans. The former is referred to as the "ventral stream" and the latter as the "dorsal stream" (Ungerleider and Haxby 1994). Lateral prefrontal areas are thought to store contextual information of the present and recent past, which can serve as a reference framework for the behavioral relevance of visual stimuli and motor plans, and can form a basis for decision-making processes (Leon and Shadlen 1998).

All of these different representations held in different cortical areas need to be integrated to form a coherent stream of perception, cognition, and action. Instead of a brain area with central executive functions, there is a massive recurrent connectivity between cortical brain areas. These connections form the white matter, which occupies the largest fraction of the brain volume. It is hypothesized

[1]Institució Catalana de Recerca i Estudis Avançats (ICREA), Barcelona, Spain
[2]Universitat Pompeu Fabra, Department of Technology Computational Neuroscience, Passeig de Circumval.lació 8, 08003 Barcelona, Spain
[3]Siemens AG, Corporate Technology, Information & Communications, 81739 Munich, Germany
[4]Department of Computer Science, Technical University of Munich, 85747 Garching, Germany

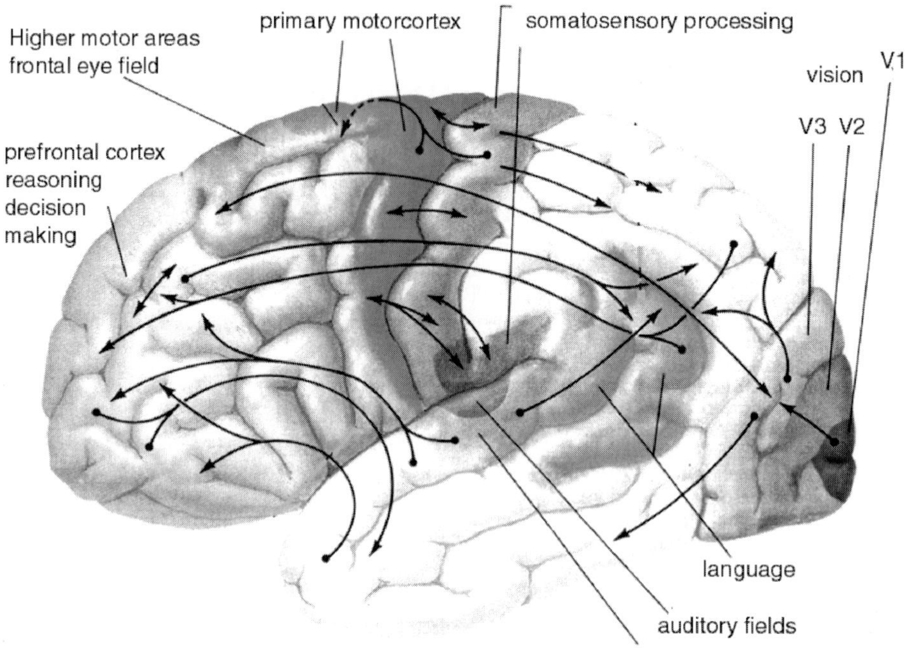

FIG. 1. Illustration of the human neocortex. Gray-shaded regions group the cortical areas by functional similarity, black arrows schematically indicate inter-areal connectivity. Adapted from Statter (2002)

that only one-quarter of all possible connections between areas have been realized in the human brain, and most of these are of recurrent nature (Salin and Bullier 1995). Thus, partial representations held in different cortical areas might be integrated by mutual cross talk, mediated by inter-areal neural fibers. Whenever one brain area provides bottom-up input to another area via inter-areal connections, the latter area feeds back top-down biasing signals, presumably to facilitate matching of the two different representations.

Further neurophysiological evidence gives rise to the assumption that each cortical area is capable of representing a set of alternative hypotheses encoded in the activities of alternative cell assemblies. Representations of different conflicting hypotheses inside each area *compete* with each other for activity and for being represented (Desimone and Duncan 1995). However, each area represents only part of the environment and / or internal state. In order to arrive at a coherent global representation, different cortical areas *bias* each others' internal representations by communicating, through inter-areal connections, their current state to other areas, thereby favoring certain sets of local hypotheses over others. For example, different objects present in the visual field could compete for being represented in one brain area. This competition might be resolved by a bias given towards one of representation from another area, as obtained from this other area's local view – encoding for example the behaviorally relevant location in the visual field and favoring thus only the object corresponding to that location to be represented in the first area (Rolls and Deco 2002). By recurrently biasing each other's competitive internal dynamics, the global neocortical system dynamically arrives at a global representation in which each area's state is maximally consistent with those of the other areas. This view has been referred to as the *biased competition hypothesis* (Moran and Desimone 1985; Chelazzi et al. 1993; Desimone and Duncan 1995; Chelazzi 1998; Reynolds and Desimone 1999).

In parallel to this competition-centered view, a *cooperation*-centered picture of brain operation has been formulated, where global representations find their neural correlate in assemblies of co-activated neurons (Hebb 1949). Co-activation of neurons induces stronger mutual connections between neurons, which lead to assembly formation. The concept of neural assemblies was later formalized in the framework of statistical physics (Hopfield 1982; Amit et al. 1994; Amit and Brunel 1997b), where assemblies of co-activated neurons form attractors in the phase space of the recurrent neural dynamics (patterns of co-activation can represent fixed points to which the dynamical system evolves). For biologically plausible networks of spiking neurons used in this study, the attractor dynamics have been recently investigated by (Amit and Brunel, 1997a; Brunel and Wang 2001; Stetter 2002; Deco and Rolls 2003).

In this chapter, we introduce the unifying principle of *biased competition and cooperation* (BCC) for neurocognitive modeling of higher neocortical functions. Section 2 presents the BCC modeling framework by summarizing a set of underlying working hypotheses and relating these hypotheses to experimental evidence. Section 3 summarizes a neurocognitive model study of attentional filtering. It shows how biased competition and cooperation operate within a single model brain area. Section 4, finally, introduces a bi-areal BCC model for learning visual categorization. It demonstrates how BCC operates across two different brain areas and shows how Hebbian synaptic plasticity can change the multi-areal attractor dynamics towards increased performance of the multi-areal system.

2 Biased Competition and Cooperation Models

2.1 Coupled Attractor Network View

The most dominant feature of the neocortex is the dense and recurrent intra-areal and inter-areal connectivity. At present, there are no clear data-derived criteria related to signal propagation time, synaptic transmission efficacy, or axonal penetrance of the target tissue that would allow clear separation of intra-areal from inter-areal connectivity. Hence, there are two alternative conceptual models for neocortical operation in the framework of recurrent network theory: (i) The first model considers the whole neocortex as a giant attractor network; its connectivity is determined by the neuroanatomical features of both the intra- and inter-areal connections. (ii) The second model treats each cortical area or even smaller sub-structures (such as a hypercolumn) as an attractor network. These smaller attractor networks are linked by recurrent long-range inter-areal connections. By these latter connections, the local attractor dynamics become linked to each other, and affect each other in such a way that a global attractor is finally formed. Because of the anatomical and functional subdivision of the neo-cortex, it seems more reasonable to adopt the second view of linked attractor networks for large-scale brain modeling. The modular architecture has the advantage that it reduces the model complexity and facilitates exploratory research.

2.2 Structural Aspects of Model Brain Area

Despite the high functional diversity, different cortical areas are remarkably uniform in their anatomical structure (Kandel et al. 1991). About 80% of neurons are excitatory pyramidal neurons (Abeles 1991), that communicate via glutamatergic AMPA (*alpha*-amino-3-hydroxy-5-methyl-4-isoxazolepropionic acid) and NMDA (N-methyl-D-aspartate) synapses. These neurons locally collect signals over a large fraction of cortical depth and laterally spread dense local excitation across a diameter of about 200 μm. Longer-range collateral axon fibers laterally spread excitation up to several millimeters, dependent on the species. A very constant feature across different areas and species is their patchy appearance (Lund et al. 1994; Bosking et al. 1997; Kisvarday et al. 1997; Somogyi et al. 1998), when viewed from the cortical surface. These patches seem to preferentially link the neurons in one area to neuron populations with similar response properties (Malach et al. 1993; Kisvarday et al. 1997). Pyramidal neurons are also the source of long-range inter-areal connectivity. A smaller amount of about 20% of cortical neurons are GABA-ergic (gamma-aminobutyric acid) and inhibitory in effect. They are highly diverse in morphology, but one prominent type of GABAergic neurons seem to be basket cells, which laterally spread inhibition through about 600–800 μm. GABAergic neurons do not directly communicate across areas (for further details see Stetter 2002, and references therein). To properly describe the dynamic aspects of neural cognitive processes, we constructed the BCC models as networks of integrated and firing neurons with detailed synaptic dynamics (as introduced by Brunel and Wang 2001). The recurrent excitatory postsynaptic currents (EPSCs) are modeled to have two components, mediated by AMPA (fast) and NMDA (slow) receptors. External EPSCs imposed onto the network from outside are assumed to be driven only by AMPA receptors. The shunting inhibitory GABAergic synapses inject inhibitory PSCs (IPSCs) into both pyramidal cells and interneurons. Furthermore, in these Models, we maintained the proportion 80% excitatory neurons and 20% inhibitory neurons, consistent with experimental data (Abeles 1991).

Motivated by the observation of cortical columns in the striate cortex, we hypothesize that cortical neurons can be grouped by the similarity of inter-areal and local input. Following the concept of population coding we adopted a spiking network structured into distinct populations of neurons. Three types of populations are defined: a specific population gathers excitatory neurons having a

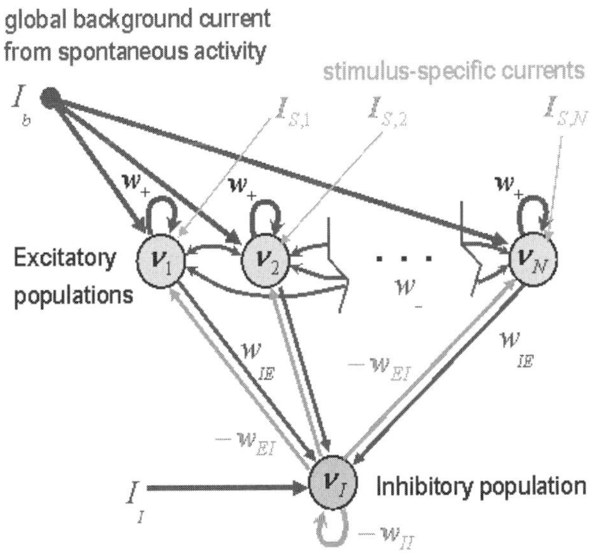

FIG. 2. Sketch of a general purpose model cortical area

specific behavioral function; a non-specific population groups all other excitatory neurons in the modeled brain area; and an inhibitory population groups all local inhibitory neurons in the modeled brain area. The latter regulates the overall activity and implements competition in the network by spreading a global inhibition signal. Within each population, neurons are mutually connected by stronger than average synaptic weights with a mean strength $w+$ (Fig. 2). These correspond to local pyramidal axonal fibers. Different populations i and j are laterally connected by weaker than average connections with mean synaptic strengths w_{ij}. The collection of all weights determines the attractor landscape and the function carried out by the model. We then introduce the following simplifying assumption, which is convenient but not a necessary ingredient to the model: Populations that represent features associated with each other are linked by stronger than average weights, $w_{ij} = w_0$. The strengthening could be the result of coactivation followed by Hebbian learning. On stimulation with one of the features, the corresponding associated populations tend to be co-activated through the recurrent intra-areal dynamics. Thus, the weights w_0 implement *cooperation* and underlie the formation of Hebbian cell assemblies in the model. However, populations that represent unrelated or anticorrelated features, are linked by weaker than average weights, $w_{ij} = w-$. The dominant connectivity between such populations is propagated laterally through the model GABAergic neurons and is inhibitory in effect. Neuron populations for different cell assemblies attempt to shut down each other's activity. Thus, the weak weights $w-$ implement *competition* for activation.

2.3 Inter-Areal Connectivity

Fast myelinated long-range axons of pyramidal neurons connect different cortical areas. They connect to spatially restricted parts of the target-area and follow some topographic order (Zeki and Shipp 1988). In most of the cases, feedforward connectivity to a target area is complemented by feedback-connectivity to the original one. The neurons feeding back from a higher area preferentially address neurons in the lower area that drives them. When an area receives input from a lower area characterized by a less abstract representation, the input is referred to as *bottom-up* driving input. Feedback input from a higher area, characterized by a more abstract representation, is referred to as *top-down* biasing input. Whereas bottom-up input is thought to activate a set of "hypotheses" consistent with the lower level (e.g., sensory) features, top-down biasing input is thought to back-propagate higher order (e.g., more global) information and thereby to contribute the selection of one activation pattern among several possible patterns.

However, although we conceptually follow this view, there is no anatomic dynamic difference between bottom-up and top-down signals in our proposed model: both form small, additive input to a given cortical area from other areas. As a consequence, a multi-areal biased competition and cooperation model consists of a recurrent network of recurrent attractor networks.

2.4 Dynamic Operation

In most cortical areas and at any time, about 99% of neurons are on average only spontaneously active at a rate of about 3 Hz (Wilson et al. 1994; Koch and Fuster 1989). About 1% of neurons are on average active with higher than spontaneous rates, typically some tens of Hz. Based on these numbers it becomes obvious that each area is mostly driven by strong background current from the ocean of spontaneously active neurons throughout the neocortex. Specific input currents are only small perturbations on top of this background current, in the range of a few percent. Hence it is the task of the recurrent areal circuitry to amplify these small inputs in a way that is useful for signal processing. Finally, cortical spike dynamics are very irregular, introducing considerable fluctuations to the synaptic currents by which the neurons communicate.

In the presence of fluctuations, intra-areal attractor dynamics can be very volatile, and can respond in dramatically different ways to small changes in driving or biasing inputs. It might be that this volatility and potential instability underlies important cognitive processes such as decision making, spontaneous thoughts and creativity.

3 Attentional Filtering

Selective attention may be defined as a process, in which the perception of certain stimuli in the environment is enhanced relative to other concurrent stimuli of less importance. A remarkable phenomenon of selective attention, known as inattentional blindness, has been described for human vision (for a review see Simons 2000). The inattentional blindness refers to an absence of awareness regarding a certain visual event when attention is focused on another event.

Recently, Everling et al. (2002), investigated the underlying mechanisms of the referred effect by measuring the activity level of the prefrontal cortex (PFC) neurons in awake behaving monkeys performing a focused attention task. In this experiment, a monkey, after being cued to attend one of two visual hemifields (left or right eye-field), had to watch a series of visual stimuli conjointly exposed in both hemifields consisting of different pairs of objects. The animal was to react with a saccade (rapid intermittent eye movement occurring when eyes fix on one point after another) if and only if a predefined target object appeared in the cued hemifield. In order to correctly perform this cognitive task, the monkey had to ignore any object in the uncued hemifield and to concentrate (focus his attention) on the cued location. The experimental results showed that some PFC neurons discriminate between a previously learned target and a non-target, but that this discrimination disappears when objects are presented in the unattended visual hemifield. We refer to this effect as attentional filtering. In other words, attention acts in a multiplicative way upon the sensory driven neuronal response, and consequently these neurons seem to code for behavioral relevance of a stimulus rather than for its identity. Only a task-relevant stimulus (i.e., target in the cued hemifield) is gated by the context and allowed to be represented. This attentional filtering effect of an object's representation for the unattended hemifield is complete and might be the neuronal substrate of the referred selective attention effect studied in humans, possibly explaining blindness to ignored inputs.

Neurodynamical models developed within the framework introduced in the second section, have been proven to successfully account for different aspects of visual attention (Rolls and Deco 2002; Corchs et al. 2003) and working memory context-dependent tasks (Deco and Rolls 2003; Deco et al. 2004; Almeida et al. 2004). Here, we review a biologically relevant minimal model (Szabo et al. 2004) for analyzing the underlying neuronal substrate of the visual attentional filtering effect. We observed that the mechanism of biased competition alone cannot account for the experimental results and show that biased competition and cooperation between stimulus selective neurons are, in combination, required conditions for reproducing the referred effect.

We implemented a network of excitatory and inhibitory integrate-and-fire neurons, modeling a small part of the PFC, which are fully connected (Fig. 3). The model (Fig. 3) consists of populations of neurons that show the same selectivities as found in the experimental results (Everling et al. 2002). Under a non-attentive control task, they encode information about the object identity ("T" for target, "O" for other) and spatial location ("L" for left, "R" for right hemifield). Therefore, we showed four interconnected selective populations coding for target with preferred location left (TL), target with preferred location right (TR), non-target (other) left (OL) and non-target (other) right (OR).

On top of the spontaneous background input received by each neuron in the network, the four selective populations are driven by object-specific and

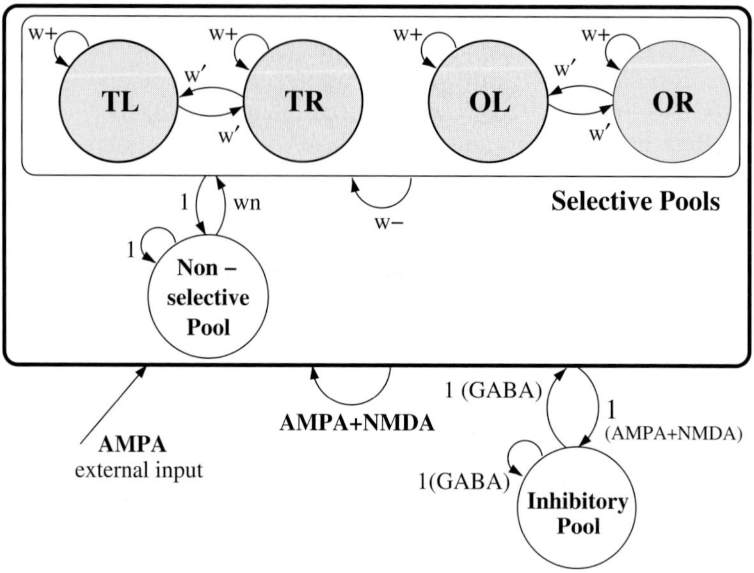

FIG. 3. Architecture of the prefrontal cortical module. The four sensory populations correspond to target and non-target selective neurons with a preferred location left or right. Adapted from Szabo et al. (2004)

unilateral inputs, assumed to originate from lower sensory areas which process the visual scene to provide these signals. Besides the specific afferent bottom-up input, the selective populations are also biased by two kinds of top down inputs. The first top-down signal biases neurons that are selective for the target object. The origin of this signal is not explicitly modeled, but it might originate from a working-memory module that encodes and memorizes context in terms of rules. The second top-down signal, the attention bias, facilitates neurons that have the cued location as a preferred location. The origin of this bias, which might be sent from a spatial working memory area, is not modeled explicitly here. The network is fully connected, but weights can differ depending on the populations being connected. We model the prefrontal cortex of a monkey that has already been trained and do not explicitly model the learning process itself. The weights between the populations were intuitively chosen such as to match Hebbian learning. Between the populations encoding the same object identity, cooperation is implemented through stronger than average weight (w'). Competition is implemented through a smaller than average weight ($w-$), as depicted in Figure 3. For more details on network implementation and parameters, see Szabo et al. (2005a).

Explicit simulations were carried out in the framework of the architecture presented in Figure 3, by applying each of the four different stimulus combinations used in Everling et al. (2002) and calculating the population-averaged spike rate of the target specific right preferred TR population. Under this condition, the attention bias set to the right preferred neurons corresponds to the condition "preferred location attended", a left bias corresponds to the "non-preferred location attended" condition. Simulation results are presented in Figure 4 (columns 2–4).

The left column of Figure 4 (Fig. 4, column 1) displays the experimental measurements recorded from the PFC of awake behaving monkeys (Everling et al. 2002) in the case of four stimulus combinations illustrated as insets. The black lines correspond to attention directed to the preferred location and the grey lines

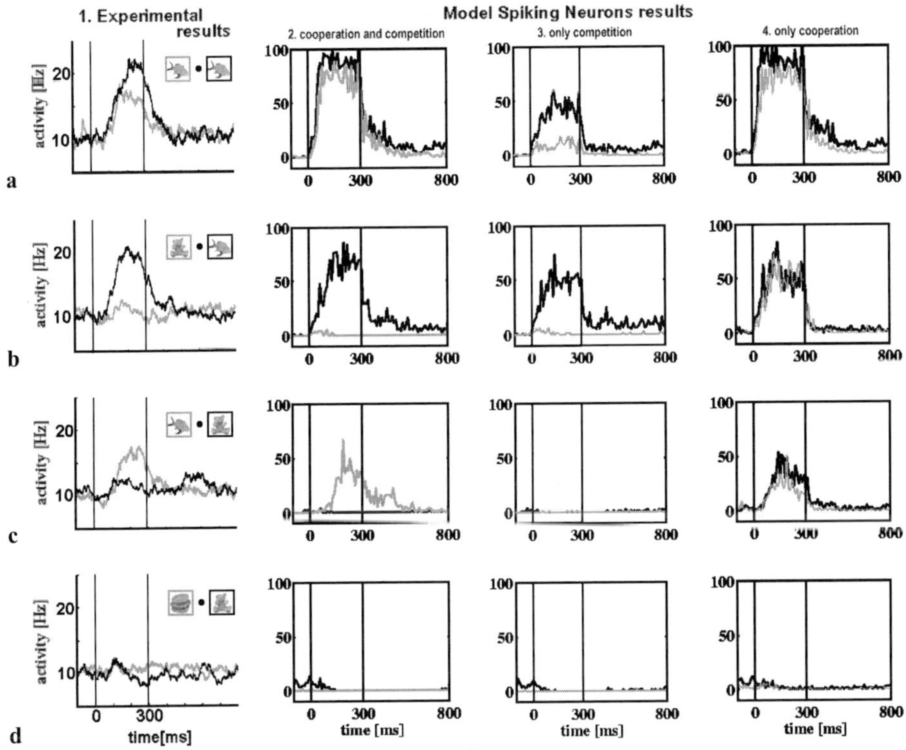

FIG. 4. Experimental results (column 1) and model simulation (columns 2–4) for focused attention task. *Black lines*: Attention focused to the preferred location (right), *grey lines*: attention focused to the non-preferred location of the measured neurons and model-neurons, respectively. **a** Both target stimuli. **b** Target in preferred location only. **c** Target in non-preferred location. **d** both non-target stimuli. Column 2: simulation with cooperation and competition . Column 3: simulation with competition only. Column 4: simulation with cooperation only. Adapted from Szabo et al. (2004)

correspond to attention directed to the non-preferred location. In the second from left column (Fig. 4, column 2), the population-averaged responses of the model "target right selective" (TR) neurons for the same stimulus conditions and attentional states as the experimental results are shown, using both mechanisms of biased competition and cooperation. From the simulation results (Fig. 4, column 2) it can be observed that with this simple network, the obtained attentional filtering effect is the same as that in the experimental results (Fig. 4, column 1).

Attentional filtering consists of four different phenomena which can be assigned to the four stimulus conditions: (i) When both hemifields contain target stimuli, the response reflects whether the attended stimulus is in the preferred or non-preferred location (Fig. 4, column 1a, column 2a). (ii) When a target appears in the preferred location only, the response is completely shut down (gray line), as soon as attention is shifted away from the target-stimulated side (Fig. 4, column 1b, column 2b). We refer to this effect as attentional suppression. (iii) In contrast, when a target appears in the non-preferred location, the neural response is increased (gray line), as soon as attention is shifted towards it (Fig. 4, column 1c, column 2c). We refer to this effect as attentional facilitation. (iv) Finally, when both hemifields are stimulated with non-targets, the response remains low, reflecting the target-selectivity of the neurons (Fig. 4, column 1d, column 2d). In combination of these effects, the neurons in both the experiment and the model encode only the contents of the attended hemifield (compare black lines in Fig. 4, column 1, column 2 a and b with c and d, compare the grey lines in Fig. 4, column 1, column 2 a and c with b and d) and ignore the contents of the non-attended hemifield (compare black lines in Fig. 4, column 1, column 2 a with b and c with d, compare the grey lines in Fig. 4, column 1, column 2 a with c and b with d). The content of the non-attended hemifield is not encoded in the responses.

When the network is dominated by competition (Fig. 4, column 3), the competition causes complete attentional suppression of unattended stimuli (Fig. 4, column 3b), however, there is no attentional facilitation (see the zero activity in Fig. 4, column 3c). This is the case, because in the present model the facilitation effect is caused by a lateral propagation of activity from the stimulated TL population to the nonstimulated TR population over recurrent connections. Because these connections are too weak in the competition only setting (i.e., w' is too small), facilitation does not occur. When the network is dominated by cooperation (Fig. 4, column 4), activities between attended and non-attended conditions are equalized, and as a consequence attentional effects are diminished (compare black with grey lines in Fig. 4, column 4). In particular, attentional suppression is no longer observed.

In summary, competition, mediated by a small weight $w-$, implements attentional suppression, and cooperation, mediated by a strong weight w', implements attentional facilitation. When both mechanisms act together, our model shows a strong, all-or-none attentional filtering effect, which results from the effects of weak top-down biases.

4 Learning to Attend

In a recent experiment performed on behaving monkeys, Sigala and Logothetis have studied how selectivity to stimulus features of infero-temporal cortical (ITC) neurons is affected by learning a visual categorization task (Sigala and Logothetis 2002). The visual stimuli (schematic images of faces, see Fig. 5 bottom-right) were characterized by several features (eye height, eye separation, nose length and mouth height), and only some of these (eye height and eye

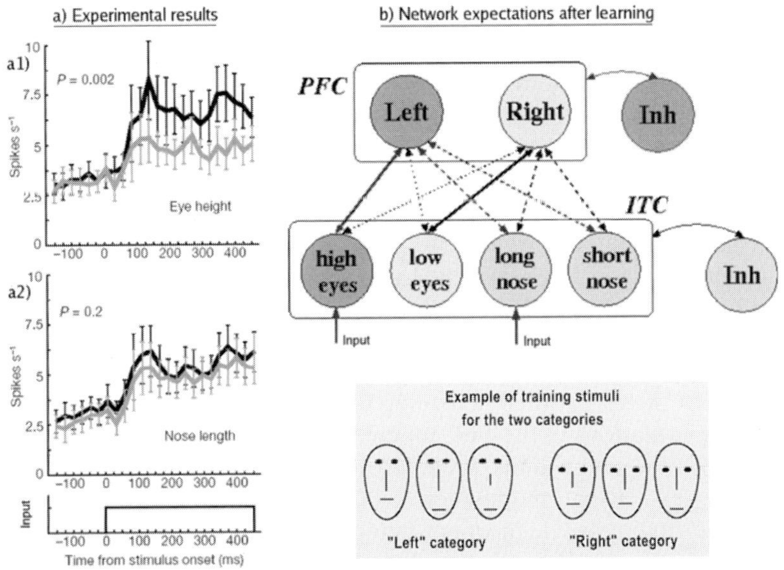

FIG. 5. **a** Experimental results adapted from Sigala and Logothetis when different combinations of features were presented (Sigala and Logothetis 2002). Shown are the average spiking rates of all recorded visually responsive neurons, grouped according to their best *(black lines)* and worst *(gray lines)* responses to the levels of diagnostic feature "Eye height" (a1) and non-diagnostic feature "Nose length" (a2). **b** Schematic representation of the network architecture and the expectations after successful learning of the visual categorization task. The connections between the diagnostic populations and the corresponding categories are potentiated *(thick arrows)*, the connections between the diagnostic populations and the non-corresponding categories are depressed *(dotted arrows)*, and the connections to and from the non-diagnostic neurons remain at an intermediate value *(dashed arrows)*. Network activities for the particular stimulus presentation characterized by "high eyes" and "long nose" are depicted by the gray levels of the populations (*dark gray*: high activity; *light gray*: low activity). The relevant information that the presented stimulus has "high eyes" will bias, through the feed-forward interlayer connections, the competition in the category model layer towards the "Left" population. This population, in turn, generates through feedback interlayer connections, the tuning of the diagnostic feature "Eye height". The categorization process does not influence the tuning of the non-diagnostic feature

separation – named diagnostic features) were relevant for the categorization task.

The experimental results showed an enhancement in neuronal tuning for the values of the diagnostic features (Fig. 5a, top). Responses to non-diagnostic features, in contrast, were poorly tuned (Fig. 5a, middle). Hence ITC activity not only encodes the presence and properties of visual stimuli but is also tuned to their behavioral relevance.

Recent studies (Freedman et al. 2003; Tomita et al. 1999) suggested that top-down signals from PFC to ITC might influence neuronal responses in ITC. Szabo et al. (M. Szabo et al., 2005) hypothesized that neuronal responses in ITC could be modulated, in a behavioral context, by top-down signals originating from category encoding neurons, possibly residing in the prefrontal cortex, PFC. They proposed a two-layer neurodynamic computational model developed in a framework of biased competition and cooperation.

The model predicted the interaction of two small connected areas in the brain, thus characterizing the stimulus-esponsive units from the ITC and the category-encoding neurons from the PFC that we will review in this section. The schematic architecture is presented in Figure 5b.

In this minimal model, it is assumed that the presented stimuli are characterized by only two features, "Eye height" and "Nose length", each with two discrete values, and that the two categories are determined exclusively only by one feature: the diagnostic feature "Eye height". Thus, there are four specific populations in the ITC layer, denoted according to the specific input that they receive. The specific populations in the PFC model layer encode two learned categories associated with the two actions: press left lever ("Left" population, or C1) and press right lever ("Right" population, or C2). The stimuli with the diagnostic feature in the first state, "high eyes", belong to category 1 and the those with diagnostic feature in the second state, "low eyes", belong to category 2, irrespective of the value of the non-diagnostic feature "Nose length".

Each individual neuron is driven by a background external input. The neurons in the four specific populations from the ITC layer additionally receive external inputs encoding stimulus specific information assumed to have on average the same strength. The network is fully connected within layers by excitatory and inhibitory synapses. Between the two layers, only specific neurons are fully connected by excitatory synapses.

In our approach we assume, for simplicity, that intra-layer connections are already formed, e.g., by earlier self organization mechanisms. In the ITC model layer, cooperation takes place between specific populations, implemented by uniform lateral connectivity. They encode the same type of stimulus and are differentiated only by their specific preferences to the feature values of the stimuli. The neural activity of the PFC model layer is designed to reflect the category to which the presented stimulus corresponded. Competition is implemented between the category encoding populations.

Connections between the ITC and PFC are modeled as plastic synapses. Their absolute strengths are learned using a reward-based Hebbian learning algorithm.

After every trial the synaptic weights are changed according to the resulting reward signal and pre- and post-synaptic population activities, until convergence to a stable configuration is reached. For more details on network structure, parameters and learning algorithms see (M. Szabo et al., 2005).

When a stimulus is presented to the trained network, after successful learning (as depicted in Fig. 5b), the sensory inputs (coming from lower visual processing areas) activate the ITC neurons and are propagated through feed-forward connections to the PFC. This bottom up input from ITC biases the competition between category encoding populations. The winning category influences the activity of the neurons in the ITC layer such that they become selective for some of the presented features. Thus, in contrast to the last section, the attentional biases needed to guide the competition are produced autonomously in the model.

Simulation results presented in Figure 6 depict average network activities (over 50 consecutive trials) in three moments of the learning process: at the beginning of learning, at an intermediate point (after 200 trials) and after the convergence of synaptic parameters following 1500 trials. The plots in the first row were obtained by performing the same calculations as for the experimental data (Fig. 5a). For each specific neuron in the ITC model layer, the spiking rates for all 50 consecutive trials were grouped based on the presented stimulus values and were averaged. Each specific neuron has a different response level to the two values of each feature. The highest responses for the diagnostic feature of all specific neurons in the ITC model area were averaged producing the "best Diagnostic" response. The lowest responses for the diagnostic feature of all specific neurons in the ITC model area were averaged to generate the "worst Diagnostic" response. Similar calculations were done for the non-diagnostic feature.

These average activities over all ITC specific neurons are presented for three points in time in Figure 6, top row. At the beginning of learning, there is no bias in the input to the PFC layer, the "Left" (C1) and "Right" (C2) populations are activated randomly with the same probability (Fig. 6a, bottom). Thus there is no difference between the tuning of the diagnostic and non-diagnostic features (Fig. 6a, top). As learning progresses and the synaptic weights evolve, the network now correctly resolves the categorization task (Fig. 6b, bottom). At the same time, we notice the beginning of the tuning process that will be enhanced in time (Fig. 6b, top). After convergence, selectivity for the level of the diagnostic feature is enhanced, as compared to the non-diagnostic feature (Fig. 6c, top). The activities for the best and worst diagnostic feature values are more separated than those for the best and worst non-diagnostic feature values. This result is in good qualitative agreement with the experimental results (Fig. 5a).

The middle and bottom rows in Figure 6 show average spiking rates of specific populations in two layers for selected trials among the 50 successive trials where the presented stimulus was characterized by "low eyes" and "long nose" (populations D2 and O1 stimulated). Since there is no structure in the model ITC layer, enhancement of selectivity emerges due to the top-down input from the PFC layer, which encodes the previously learned stimulus categories. The rightmost

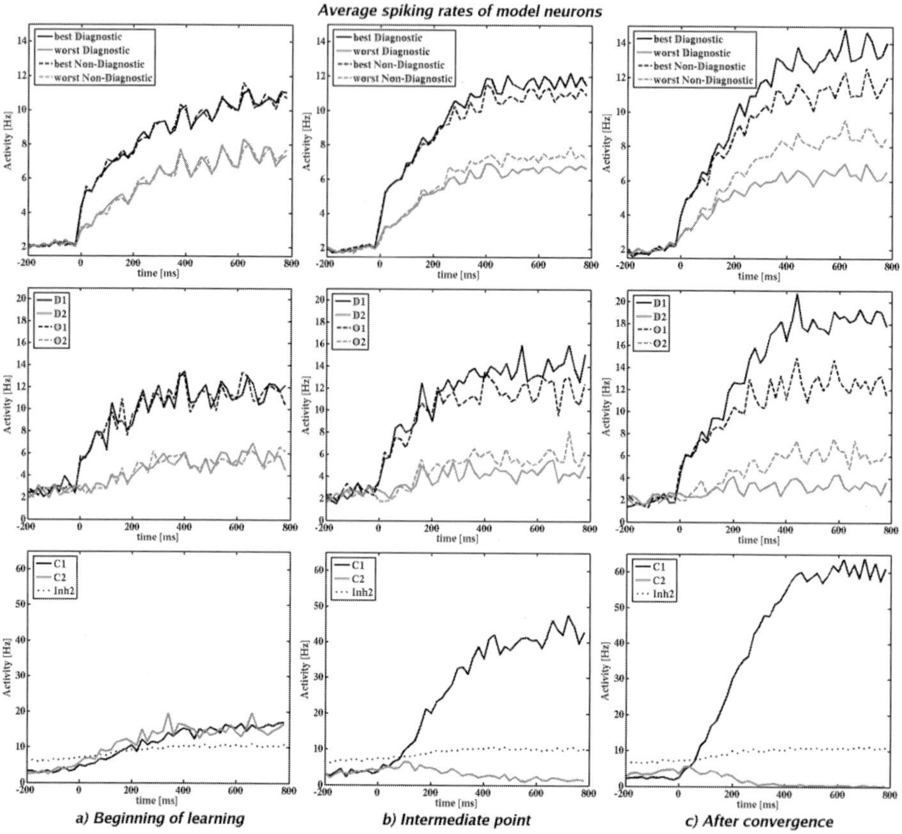

FIG. 6. Simulation results for a spiking network averaged over 50 successive trials at three points in the learning process: **a** at the beginning of learning; **b** an intermediate point during learning (after 200 steps); **c** after the weights converged to a stable configuration (1,500 steps). The top row shows average spiking rates of stimulus responsive neurons, grouped according to their best and worst responses to the levels of diagnostic and non-diagnostic features. The middle and bottom rows show the average spiking rates of specific populations in the ITC layer (D1, D2, O1, O2) and the PFC layer (C1, C2), respectively, for trials where the presented stimulus was characterized by: low eyes and long nose (external input to the populations D2 and O1) among 50 successive trials. Adapted from Szabo et al. (2006)

column, Figure 6c, corresponding to the point in the learning process, where the weights converged to a stable configuration, is in agreement with the expectations after learning depicted in Figure 5b. From the time when the stimulus is presented to the network (time = 0 ms in Fig. 6), the selectivity of the category specific populations (Fig. 6c, bottom row) emerges through the competition biased by feed-forward inputs (ITC → PFC) from the specific populations of the ITC layer. Through the feedback modulatory inputs (PFC → ITC), this selectivity is transmitted afterwards to the feature-specific populations in the ITC (Fig. 6c, middle). It can be seen that in the first 100 ms after stimulus onset, the D1 and O1 (stimulated) or D2 and O2 (non-stimulated) populations do not differ in activity. Hence there is no diagnostic tuning. Only after the correct category population acquires activity, the diagnostic tuning builds up.

Summarizing the results of our simulations, we consider that the enhancement of selectivity for behaviorally relevant features could result from a constructed reward-based Hebbian learning scheme. The latter scheme robustly modifies the connections between the feature encoding layer (ITC) and the category encoding layer (PFC) to a setting where the neurons activated by the level of a feature determinant for categorization are strongly connected to the associated category and weakly connected to the other category, and the neurons that receive input specific for a task-irrelevant feature, are connected to the category neurons with an average weight, not significantly changed during training. In summary, the network successfully develops both a forward IT→PFC synaptic structure able to support correct classification, and a backward PFC→IT synaptic structure producing a task-dependent modulation of IT response, providing evidence of a qualitative agreement with the findings of Sigala and Logothetis.

References

Abeles A (1991) Corticonics. Cambridge University Press, New York

Almeida R, Deco G, Stetter M (2004) Modular biased-competition and cooperation: a candidate mechanism for selective working memory. Eur J Neurosci 20(10):2789–2803

Amit DJ, Brunel N (1997a) Dynamics of a recurrent network of spiking neurons before and following learning. Network Comput Neural Syst 8:373–404

Amit DJ, Brunel N (1997b) Model of global spontaneous activity and local structured (learned) delay activity during delay periods in cerebral cortex. Cereb Cortex 7:237–252

Amit DJ, Brunel N, Tsodyks M (1994) Correlations of cortical hebbian reverberations: experiment versus theory. J Neurosci 14:6435–6445

Bosking WH, Zhang Y, Schofield B, Fitzpatrick D (1997) Orientation selectivity and the arrangement of horizontal connections in tree shrew striate cortex. J Neurosci 17: 2112–2127

Brunel N, Wang XJ (2001) Effects of neuromodulation in a cortical network model of object working memory dominated by recurrent inhibition. Comput Neurosci 11:63–85

Chelazzi L (1998) Serial attention mechanisms in visual search: a critical look at the evidence. Psychol Res 62:195–219

Chelazzi L, Miller E, Duncan J, Desimone R (1993) A neural basis for visual search in inferior temporal cortex. Nature 363:345–347

Corchs S, Stetter M, Deco G (2003) System-level neuronal modeling of visual attentional mechanisms. Neuroimage 20:143–160

Deco G, Rolls ET (2003) Attention and working memory: a dynamical model of neuronal activity in the prefrontal cortex. Eur J Neurosci 18:2374–2390

Deco G, Rolls ET, Horwitz B (2004) "What" and "where" in visual working memory: a computational neurodynamical perspective for integrating fmri and single-neuron data. J Cogn Neurosci 16:683–701

Desimone R, Duncan J (1995) Neural mechanisms of selective visual attention. Annu Rev Neurosci 18:193–222
Everling S, Tinsley C, Gaffan D, Duncan J (2002) Filtering of neural signals by focused attention in the monkey prefrontal cortex. Nat Neurosci 5:671–676
Freedman DJ, Riesenhuber M, Poggio T, Miller EK (2003) A comparison of primate prefrontal and inferior temporal cortices during visual categorization. J Neurosci 23:5235–5246
Hebb DO (1949) The organization of behavior. Wiley, New York
Hopfield JJ (1982) Neural networks and physical systems with emergent collective computational abilities. Proc Natl Acad Sci U S A 79:2554–2558
Kandel ER, Schwartz JH, Jessel TM (1991) Principles of neural sciences. Prentice Hall International, London.
Kisvarday ZF, Toth E, Rausch M, Eysel UT (1997) Orientationspecific relationship between populations of excitatory and inhibitory lateral connections in the visual cortex of the cat. Cereb Cortex 7:605–618
Koch KW, Fuster JM (1989) Unit activity in monkey parietal cortex related to haptic perception and temporary memory. Exp Brain Res 76:292–306
Leon ML, Shadlen MN (1998) Exploring the neurophysiology of decisions. Neuron 21:669–672
Lund JS, Levitt JB, Wu Q (1994) Topography of excitatory and inhibitory connectional anatomy in monkey visual cortex. In: Lawton TB (Ed) Computational vision based on neurobiology. SPIE, Bellingham WA, pp 174–184
Malach R, Amir Y, Harel M, Grinvald A (1993) Relationship between intrinsic connections and functional architecture revealed by optical imaging and in vivo targeted biocytin injections in primate striate cortex. Proc Natl Acad Sci USA 90:10469–10473
Moran J, Desimone R (1985) Selective attention gates visual processing in the extrastriate cortex. Science 229:782–784
Reynolds J, Desimone R (1999) The role of neural mechanisms of attention in solving the binding problem. Neuron 24:19–29
Rolls ET, Deco G (2002) Computational neuroscience of vision. Oxford University Press, Oxford
Salin P, Bullier J (1995) Corticocortical connections in the visual system: structure and function. Physiol Rev 75:107–154
Sigala N, Logothetis N (2002) Visual categorization shapes feature selectivity in the primate temporal cortex. Nature 415:318–320
Simons DJ (2000) Attentional capture and inattentional blindness. Trends Cognit Sci 4:147–155
Somogyi P, Tamas G, Lujan R, Buhl EH (1998) Salient features of synaptic organization in the cerebral cortex. Brain Res Brain Res Rev 26:113–135
Stetter M (2002) Exploration of cortical function. Kluwer Academic Publishers, Dordrecht
Szabo M, Almeida R, Deco G, Stetter M (2004) Cooperation and biased competition model can explain attentional filtering in the prefrontal cortex. Eur J Neurosci 19:1969–1977
Szabo M, Almeida R, Deco G, Stetter M (2005) A neuronal model for the shaping of feature selectivity in it by visual categorization. Neurocomputing 65–66:195–201
Tomita H, Ohbayashi M, Nakahara K, Hasegawa I, Miyashita Y (1999) Top-down signal from prefrontal cortex in executive control of memory retrieval. Nature 401:699–703

Ungerleider LG, Haxby JV (1994) "What" and "where" in the human brain. Curr Opin Neurobiol 4:157–165
Van Essen DC, anderson CH, Felleman DJ (1992) Information processing in the primate visual system: an integrated systems perspective. Science 255:419–423
Wilson F, Scalaidhe S, Goldman-Rakic P (1994) Functional synergism between putative gamma-aminobutyrate-containing neurons and pyramidal neurons in prefrontal cortex. Proc Natl Acad Sci USA 91:4009–4013
Zeki S, Shipp S (1988) The functional logic of cortical connections. Nature 335:311–317

Part III
Action

13
Influence of Visual Motion on Object Localisation in Perception and Action

Hiroshi Ashida

1 Introduction

The topic of this chapter is visual localisation of objects. Object recognition normally refers to the ability to identify *what* it is without concerned for *where* it is. In other words, the question is how we obtain a location-invariant representation of object. There is also a rationale derived from physiological findings indicating two separate pathways for *what* and *where* information (Ungerleider and Mishkin 1982). However, it is often equally important in real life to know where the object lies. We cannot eat an apple if we can not reach it with our hand and grasp it. To do this, we need to know its precise location together with its identity as a fresh apple that can be eaten. Object localisation is therefore closely related to object recognition in an ecological sense, and it would make sense to take a short break from the intense discussion on recognition in this book to consider localisation.

More specifically, recent findings on the role of visual motion on spatial localisation will be discussed. We sometimes need to interact with objects that move across the visual field. This happens daily when you walk on a busy street or play with your cat, but it is more typical in sports such as baseball, cricket, and soccer in which the players need to interact with fast moving balls. Of course, we need to develop our motor skill to achieve good performance, but it is also expected that the visual system has been evolved to cope with dynamic interaction with objects.

A problem then is that neural signal processing is rather slow. For example, the latency typically measured in macaque striate cortex is about 30 to 50 ms (Maunsell and Gibson 1992). A ball coming at a speed of 150 km/h travels more than one meter during this delay, leaving no chance of hitting it. Obviously, we need to have some methods of anticipating the target path. Given that the delay in physical action is large and effector-dependent, it is likely that most of the adjustment should be accomplished through motor planning and its execution. However, the visual system seems to have its own process for delay

Graduate School of Letters, Kyoto University, Kyoto 606-8501, Japan

compensation, as suggested by several motion-related illusions. Also, such compensation might work specifically for visuomotor action without always being consciously perceived. Here, I review such illusions after brief summary of separate systems of vision for perception and action (Milner and Goodale 1995), and describe results from our group that indicated action-specific visual motion extrapolation.

2 Vision for Perception and Action

2.1 Separate Visual Pathways for Perception and Action

Milner and Goodale (1995) proposed that the brain has separate visual pathways for conscious perception and direct visuomotor control, and this proposal has been followed by intensive discussion over the past decade. They extended the idea of two visual pathways for *what* and *where* information (Ungerleider and Mishkin 1982) and argued that the ventral pathway is dedicated to the detailed conscious perception, while the dorsal pathway is dedicated to direct control of action (Fig. 1). It was radically assumed that the two pathways are independent and information through the dorsal pathway is not always accessible to conscious perception.

Supporting evidence for their theory has mainly come from case studies of human brain damage patients and lesion studies of monkeys. A patient with visual form agnosia was able to perform precise action like grasping or mailing without being able to perceive the detail (Goodale et al. 1991; Milner et al. 1991). There are also cases of "blindsight" patients who can point to the target without conscious perception (Weiskranz 1986). These patients generally have damage in the occipital lobe, and somtimes in the primary visual cortex (V1), that causes an overall dysfunction of the ventral stream. The dorsal pathway is relatively intact with possible support from the subcortical path through the superior col-

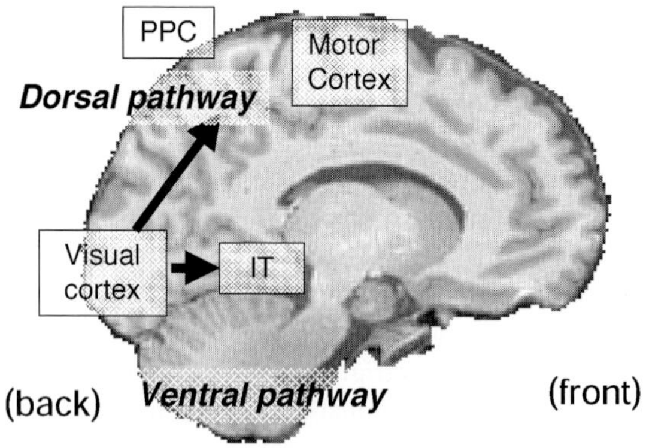

FIG. 1. The two major visual pathways in a human brain. From the visual cortex in the occipital lobe, the dorsal pathway extends into the posterior parietal cortex (*PPC*) through V5, while the ventral pathway goes into the inferotemporal (*IT*) cortex

liculus, which is considered to enable visuomotor coordination. On the other hand, patients with optic ataxia (Bálint syndrome) tend to show difficulties in visually-guided actions like reaching, while conscious perception is relatively unaffected (Bálint 1909). This syndrome generally involves damage in the parietal lobe. Recently, it has been suggested that the damage is more specific to direct or on-line visuomotor control in a specific area within the dorsal pathway (Glover 2003; Rossetti et al. 2003). Interestingly, the accuracy of pointing action was improved if the patient waited for 5s before initiating the action (Milner et al. 1999), which supports this recent view. There is insufficient space to describe the details here, but the results of monkey studies basically parallel these findings; lesion in the posterior parietal areas causes disorder in visuomotor action while lesion in the infero-temporal areas disrupts perceptual judgements (see Milner and Goodale 1995).

2.2 Psychophysics on Dissociation of Perception and Action

There are psychophysical results that suggest similar dissociation in normal human observers. Displacement of a target near the time of saccade is not noticed but pointing action can be accurately performed (Bridgeman et al. 1979). A stationary target appears to move when the surrounding frame moves back and forth, but reaching action is not affected (Bridgeman et al. 1981). Controversies have arose after Aglioti et al. (1995) reported that grasping action is not markedly affected by the size illusion of Titchener-Ebbinghaus circles (Fig. 2a). The "maximum aperture size" between the thumb and the index finger varied in relation to the actual object size, but it was relatively unaffected by the size contrast illusion induced by surrounding larger and smaller disks. They argued that the hand is not deceived by the illusion in conscious perception.

There have been criticisms of this experiment by Aglioti et al. (Franz 2001; see also Carey 2001). The most controversial point was the different task requirements. The perceptual task inherently involved comparison of two central disks, but the grasping task did not require this once the participant decided which target to pick. When the figure was shown one by one in both cases, there was no difference between perception and action (Franz et al. 2000). It has also been pointed out that the use of reference frames might cause different results (Bruno

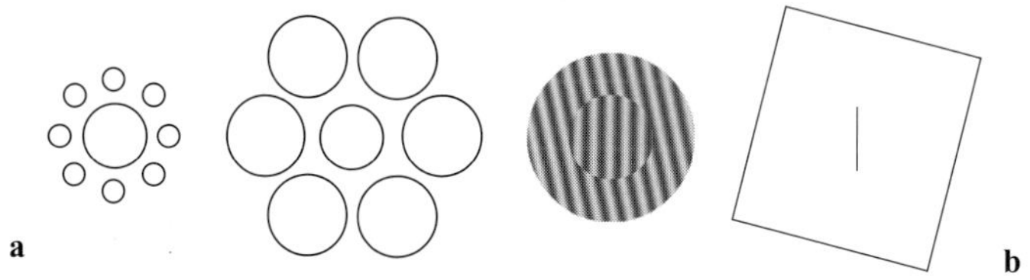

FIG. 2. **a** The size contrast illusion (Tichener-Ebbinghaus circles). The left central disk appears larger than the right one because of the surrounds, although they are of the same size. **b** The orientation contrast effect used by Dyde and Milner (2002). The central grating on the left appears tilted clockwise due to the adjacent tilted grating. The central vertical line in the right panel appears tilted counter-clockwise due to the tilted frame

2001). It seems that we have not come to a final conclusion, but the criticisms also have problems. As Milner and Goodale (1995) originally pointed out, different use of reference frames could be inherent in perception and action. The conclusion of Aglioti et al. seems to make sense even with the difficulties in the task differences. Findings of Bridgeman et al. regarding manual pointing seem less controversial because the task requirement of target localisation is equal in perception or action.

But why is action necessarily more accurate than perception? A case where action is more erroneous would complete double dissociation. Dyde and Milner (2002) conducted a clever experiment in which the illusion is cancelled for perception but not for action, leading to apparent larger errors in action. First, they showed that orientation contrast between adjacent gratings (as seen in Fig. 2b, left) affected both perception and action (mailing: to orient a card as if putting it between bars) while a far frame induces orientation contrast only in perception (Fig. 2b, right). Then, when the left pattern is surrounded by an oppositely-tilted frame, the perceptual effects are cancelled out, but the visuomotor effect is not. The measured illusion was actually larger in action than in perception. This result demonstrated double dissociation between perception and action when coupled with the case where perceptual errors were larger with the far frames alone. They reasoned that the contrast between adjacent gratings occurs at an early level where the two pathways have not branched, while the frame effect occurs later in the vision-for-perception pathway. Their results, however, do not provide sufficient evidence for separate visual pathways. Action might ignore visual processing at a later stage, but this does not necessarily mean that the visual information is separately elaborated for visuomotor control. A critical case is missing where vision for action is directly more susceptible to an illusion, which is fulfilled by our results on motion-related illusions.

3 Motion Extrapolation Revealed by Visual Illusions

3.1 Flash Lag

When a visual object is briefly presented near a continuously moving visual object, the moving one is perceived ahead of the flashed one (Fig. 3a), which has been called "flash lag" (FL). This effect had been already reported by MacKay (1958), but Nijhawan (1994) reinterpreted it as evidence for extrapolation of target motion to compensate for the neural delay, which has triggered numbers of follow-up studies. Nijhawan considered that compensation is particularly important for catching action, although it was just speculative.

Unfortunately, this very intriguing idea of motion extrapolation has not been supported by later studies. A major objection was that no overshoot of target motion is perceived if the moving target turns back at the time of the flash (Whitney and Murakami 1998). Extrapolation should have resulted in shift in the direction of the target motion before the unexpected reversal, but the moving

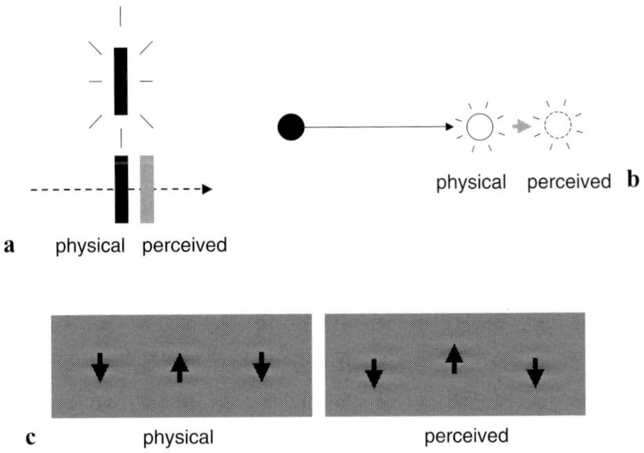

FIG. 3. **a** The flash lag illusion. A bar briefly flashes over a moving bar. When the two bars are physically aligned, we perceive that the flashed bar lags behind the moving one and they are not aligned. **b** The "representational momentum". The final location of a suddenly disappeared moving target is often perceived to be shifted ahead. **c** Motion related positional shift. When the three drifting Gabor patches with stationary windows are vertically aligned, the central patch looks misaligned, shifted in the direction of motion. These figures illustrate typical displays, but many variations have been demonstrated

target is actually perceived as shifted in the direction after the reversal (Eagleman and Sejnowski 2000; Whitney and Murakami 1998). Furthermore, while no offset is perceived when the moving target disappears with the flashed one (flash terminated cycle, FTC), clear offset is perceived when the flash appear together with the moving one (flash initiated cycle, FIC) although there is no prior motion for extrapolation (Eagleman and Sejnowski 2000). Moreover, the size of the FL depends on the motion speed after the flash (Brenner and Smeets 2000). All of these observations contradict the extrapolation hypothesis. It is rather suggested that the perceived spatial offset is caused by a delay in the processing of the flashed target. In other words, FL is caused by the latency difference between continuous and suddenly-appearing targets (Whitney and Murakami 1998; Whitney et al. 2000), as the term "flash lag" correctly implied. Reduction of latency for a moving target can be related to attention (but see also Khurana et al. 2000; Namba and Baldo 2004). But latency difference alone might not be sufficient (Arnold et al. 2003). Full explanation would include several factors like temporal averaging before and after the flash, and occasional release from it ("postdiction" by Eagleman and Sejnowski 2000).

The FL phenomenon therefore does not prove target extrapolation for "predicting the present" (Cavanagh 1997). However, a shorter latency for a moving object would at least partially compensate for the delay to facilitate action control. It is notable then that similar phenomenon has been reported cross-modally between vision and hand movement (Nijhawan and Kirschfeld 2003).

3.2 Representational Momentum

When a moving target suddenly disappears, its final position tends to be perceived ahead of the physical position (Fig. 3b). This phenomena has been called "representational momentum" (abbreviated as RM or sometimes "RepMo") since it is as if our internal representation of the target has a momentum that cannot stop immediately (Freyd and Finke 1984). Intuitively, RM is understood as a signature of visual motion extrapolation. Note that the terminology does not necessarily imply the underlying mechanism at least in this article.

RM is apparently related to the FL (flash lag), but these two are distinct with regard to whether it involves relative judgement of position. The situation of RM is similar to the FTC case of FL, but note that the general finding of no FL for FTC is therefore not contradictory with RM. Interestingly, FL can occur even in the FTC if the spatial uncertainty is increased (Kanai et al. 2004).

As the terminology suggests, RM has been considered a cognitive effect on a memorised representation of the target, as supported by the effects of gravity and surface friction (Hubbard 1995). However, the basic effect might occur at an early perceptual level. Pursuit eye movement is crucial especially for a linear motion path; when the observer maintained fixation, the effect nearly disappears (Ashida 2004; Kerzel 2000). Cognitive extrapolation could have been more stable when the visual motion is more accurately coded in the vicinity of eye movement, but it is not. Kerzel proposed that overshoot of pursuit eye movement should be the direct cause of the perceived RM, coupled with visible persistence and centrifugal bias (Kerzel 2000).

3.3 Motion-Related Positional Shift

Perception of object position is more directly affected by visual motion signals. A typical example is a drifting grating seen through a stationary window, when the edges are blurred as in Gabor patches. The whole window is perceived as shifted in the direction of the carrier motion (De Valois and De Valois 1991) so that aligned patches of oppositely drifting carriers do not appear aligned (Fig. 3c).

This illusion has been considered to reflect spatial extrapolation for compensation of neural delays (Anstis and Ramachandran 1995). Technically speaking, however, there is no need to extrapolate the position of the stationary window. This suggests that the spatial shift is caused by a simple automatic process at a relatively early level. It even does not require real retinal motion signals, because adaptation to motion causes opposite spatial shifts (Nishida and Johnston 1999; Snowden 1998) with perceived motion aftereffect (MAE) in the stationary pattern. Even visible MAE does not seem a necessary condition for positional shifts. While MAEs are selective to spatial frequency (see Mather et al. 1998 for general reviews), the positional shift was immune to it; when the carrier orientation in the test pattern was orthogonal to that of the adapting one, we see little or no MAE but still see positional shifts (McGraw et al. 2002). Conscious perception of adapting motion is not necessary, either; positional shift occurs when the adapting motion is not identifiable due to crowding (Whitney 2005). Underlying mechanisms for the positional shift are still open for further studies, but these results suggest that positional shifts reflect early internal motion signals regardless of final perception of motion.

Visual motion in the background area also affects target localisation. The position of a briefly presented target is shifted in the direction of a drifting grating even when the target is spatially separated from the grating (Whitney and Cavanagh 2000). It is as if motion stimuli distort the whole visual field, but an important difference is that the target must be presented only for a short period. The background motion probably helps to compensate for our body or eye movement in order to point to the target accurately (Whitney et al. 2003b).

3.4 Visual Motion and Reaching: Evidence of Extrapolation for Action?

Visual illusions should be related to ecological roles of specific visual functions, if they may not have obvious ecological merits themselves. In this respect, the motion-related illusions described above should be more closely related to direct action if they reflect some operations for delay compensation. Flash lag involves a relative judgement of positions that is not easily tested by action in an unbiased way, but the other two illusions have been tested in similar conditions for perception and action.

We have reported that motion-related positional shift is more prominent in open-loop reaching action than in perceptual judgement (Yamagishi et al. 2001). A Gabor patch with a drifting vertical sinusoidal carrier was presented briefly to the right of fixation. The observers then judged the horizontal location of the target and responded either by touching the location using a rubber pen (*visuomotor* task) or by reading a visual ruler that was presented on the screen (*perceptual* task). Note that the task requirement was similar and there is no task-dependent bias for different reference frames. In the visuomotor task, the observers made ballistic movement of their hand without seeing their arm and hand (open-loop action). The stimuli were observed through a mirror for this purpose. The absolute locations of responses were not always veridical without feedback, and we computed the averaged difference in responses to leftward and rightward stimuli as an index of the effect of carrier motion on localisation. The left panel of Figure 4 shows a typical result from one observer. Obviously, localisation error in the visuomotor task was larger than that in the perceptual task, increasing more rapidly with carrier speed. This difference cannot be attributed to the intrinsic open-loop gain of the motor system, because the difference in the two response modes almost disappeared when the response was delayed by 4 s (Fig. 4, right panel). Delayed responses had to rely on the stored perceptual representation (Hu and Goodale 2000; Milner et al. 1999). We also suggested that visuomotor responses are less asymmetric than perceptual ones with regard

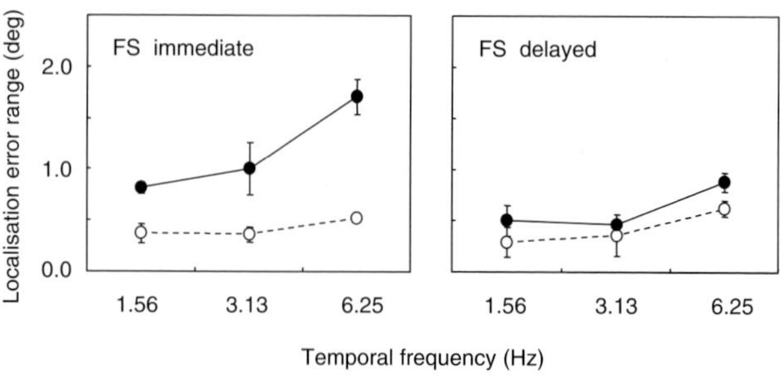

FIG. 4. Localisation errors for a drifting Gabor patch in perception and action for one observer. Differences in the mean responses for leftward and rightward stimuli are shown as a function of temporal frequency (speed). Immediate responses (*left*) and delayed (by 4s) responses (*right*). Adapted from Yamagishi et al. (2001)

to the motion direction. The result supports separate visual mechanisms, demonstrating a case where action is more prone to illusion.

We have also found that the enhanced visuomotor localisation error is specific to achromatic stimuli (Ashida et al. 2005). Equiluminant chromatic stimuli (red-green) did not yield a significant difference between perception and action. Given the weak response of V5 to chromatic stimuli (Gegenfurtner et al. 1994), it is tempting to conclude that the visuomotor-specific positional errors occur within the dorsal visual pathway, while perceptual errors reflect interaction of the two pathways where chromatic and achromatic motion signals are integrated.

A study of RM supported these findings (Ashida 2004). The final position of a horizontally moving disk on the screen was indicated using an on-screen cursor (perceptual) or by directly touching the screen (visuomotor). Visual feedback was controlled using a liquid crystal shutter goggle. The main result is shown in Figure 5, which demonstrates three major findings. First, open-loop action yielded larger forward shifts that increased with target speed more linearly than perception, which is very similar to the left panel of Figure 4. Second, closed-loop responses were almost identical to the perceptual ones. It seems that perceptual information was dominant in this case. Finally, and most interestingly, perceptual shifts were reduced to almost zero by eye fixation (Kerzel 2000), but open-loop responses remained nearly the same. This implies an intriguing possibility that extrapolation might occur within the egocentric coordinate that would be the default in visuomotor action. Perception might rely more on a retinotopic or possibly allocentric coordinate (not distinguishable under this condition). It is conjectured that perceptual RM occurs because perception uses egocentric signals when the retinotopic signals are unstable due to eye movements. In any case, further evidence was provided for separate visual processing for perception and visuomotor action in qualitative as well as quantitative ways.

While these results in general agree with the theory of Milner and Goodale (Milner and Goodale 1995), one problem arises regarding anatomical structures. They proposed that conscious perception arises only within the ventral brain

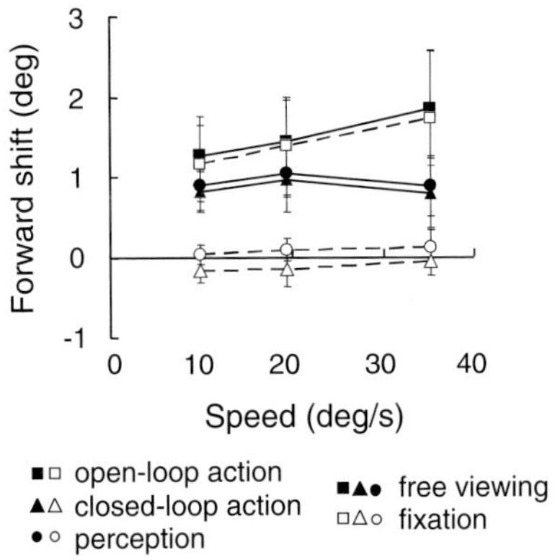

FIG. 5. The "representational momentum" for perception and action. Illusory forward shifts are plotted as a function of the target speed (n = 4, averaged). Adapted from Ashida (2004)

pathway. But if so, how can we understand conscious perception of visual motion that is believed to be based upon area V5 (MT/MST) within the dorsal pathway? The basic idea of two visual systems has been confirmed by the results, but the underlying anatomical structure should be reconsidered. It now seems more plausible to assume that some parts of the dorsal pathway are involved in conscious perception. According to Rizzolatti and Matelli (2003), there are two distinct pathways within the dorsal pathway, one from V6 to the superior parietal lobule and the other from V5 to the inferior parietal lobule. The former is considered to support on-line visuomotor control while the latter might underlie space perception (Ungerleider and Mishkin 1982). More studies would be required for further understanding of the two pathways.

4 Concluding Remarks

Effects of visual motion on spatial localisation have been extensively studied over the past several years. I have concentrated on manual action, but eye movements have also been studied for visuomotor coordination, as partly discussed in the chapter by Sogo and Osaka in this book.

However, we have not yet come to understand the underlying neural mechanism. We have been surprised by the fMRI (functional magnetic resonance imaging) results that the stimulus representation in V1 might be shifted in the opposite direction by visual motion (Whitney et al. 2003a). Although it has turned out that the effect is small and there is no overall opposite shifts (Ashida and Smith 2005; Liu et al. 2004), it is plausible that early visual areas are not responsible for motion-related shifts, which is also suggested by dissociation between perception and action in our studies. Activities in V1 would have affected both equally. Despite a positive result in cat's primary visual cortex (Fu et al.

2004), higher areas should be sought for humans as suggested by a TMS (transcranial magnetic stimulation) study (McGraw et al. 2004); giving TMS to V5 reduced the positional shifts after motion adaptation but TMS to V1 had no effect. Techniques have been developed to investigate higher and smaller cortical areas and new insights are expected to be provided in the near future.

Acknowledgement. This work was supported by the 21st Century COE progam (D-10 to Kyoto University), MEXT, Japan.

References

Aglioti S, DeSouza JF, Goodale MA (1995) Size-contrast illusions deceive the eye but not the hand. Curr Biol 5:679–685

Anstis S, Ramachandran VS (1995) At the edge of movement. In: Gregory R, Harris J, Heard P, Rose D (Eds) The artful eye. Oxford University Press, Oxford, pp 232–248

Arnold DH, Durant S, Johnston A (2003) Latency differences and the flash-lag effect. Vision Res 43:1829–1835

Ashida H (2004) Action-specific extrapolation of target motion in human visual system. Neuropsychologia 42:1515–1524

Ashida H, Yamagishi N, Anderson SJ (2005) Visually-guided actions are dependent on luminance signals. Perception 34:245

Bálint R (1909) Seelenlähmung des “Schauens”, optische Ataxie, räumliche Störung der Aufmersamkeit. Monatschrift für Psychiatrie und Neurologie 25:51–81

Brenner E, Smeets JB (2000) Motion extrapolation is not responsible for the flash-lag effect. Vision Res 40:1645–1648

Bridgeman B, Lewis S, Heit G, Nagle M (1979) Relation between cognitive and motor-oriented systems of visual position perception. J Exp Psychol Hum Percept Perform 5:692–700

Bridgeman B, Kirch M, Sperling A (1981) Segregation of cognitive and motor aspects of visual function using induced motion. Percept Psychophys 29:336–342

Bruno N (2001) When does action resist visual illusions? Trends Cogn Sci 5:379–382

Carey DP (2001) Do action systems resist visual illusions? Trends Cogn Sci 5:109–113

Cavanagh P (1997) Visual perception. Predicting the present. Nature 386:19, 21

De Valois RL, De Valois KK (1991) Vernier acuity with stationary moving gabors. Vision Res 31:1619–1626

Dyde RT, Milner AD (2002) Two illusions of perceived orientation: one fools all of the people some of the time; the other fools all of the people all of the time. Exp Brain Res 144:518–527

Eagleman DM, Sejnowski TJ (2000) Motion integration and postdiction in visual awareness. Science 287:2036–2038

Franz VH (2001) Action does not resist visual illusions. Trends Cogn Sci 5:457–459

Franz VH, Gegenfurtner KR, Bulthoff HH, Fahle M (2000) Grasping visual illusions: no evidence for a dissociation between perception and action. Psychol Sci 11:20–25

Freyd JJ, Finke RA (1984) Representational momentum. J Exp Psychol Learn Mem Cogn 10:126–132

Fu YX, Shen Y, Gao H, Dan Y (2004) Asymmetry in visual cortical circuits underlying motion-induced perceptual mislocalization. J Neurosci 24:2165–2171

Gegenfurtner KR, Kiper DC, Beusmans JMH, Carandini M, Zaldi Q, Movshon JA (1994) Chromatic properties of neurons in macaque MT. Vis Neurosci 11:455–466

Glover S (2003) Optic ataxia as a deficit specific to the on-line control of actions. Neurosci Biobehav Rev 27:447–456

Goodale MA, Milner AD, Jakobson LS, Carey DP (1991) A neurological dissociation between perceiving objects and grasping them. Nature 349:154–156

Hu Y, Goodale MA (2000) Grasping after a delay shifts size-scaling from absolute to relative metrics. J Cogn Neurosci 12:856–868

Hubbard TL (1995) Cognitive representation of motion: evidence for friction and gravity analogues. J Exp Psychol Learn Mem Cogn 21:241–254

Kanai R, Sheth BR, Shimojo S (2004) Stopping the motion and sleuthing the flash-lag effect: spatial uncertainty is the key to perceptual mislocalization. Vision Res 44:2605–2619

Kerzel D (2000) Eye movements and visible persistence explain the mislocalization of the final position of a moving target. Vision Res 40:3703–3715

Khurana B, Watanabe K, Nijhawan R (2000) The role of attention in motion extrapolation: are moving objects "corrected" or flashed objects attentionally delayed? Perception 29:675–692

Liu J, Ashida H, Smith AT, and Wandell BA (2006) Assessment of stimulus induced changes in human VI visual field maps. J Neurophysiol 96:3398–3408

Mackay DM (1958) Perceptual stability of a stroboscopically lit visual field containing self-luminous objects. Nature 181:507–508

Mather G, Verstraten FAJ, Anstis S (1998) The motion aftereffect: a modern perspective. MIT Press, Cambridge MA

Maunsell JH, Gibson JR (1992) Visual response latencies in striate cortex of the macaque monkey. J Neurophysiol 68:1332–1344

McGraw PV, Whitaker D, Skillen J, Chung ST (2002) Motion adaptation distorts perceived visual position. Curr Biol 12:2042–2047

McGraw PV, Walsh V, Barrett BT (2004) Motion-sensitive neurones in V5/MT modulate perceived spatial position. Curr Biol 14:1090–1093

Milner D, Goodale MA (1995) The visual brain in action. Oxford University Press, Oxford

Milner AD, Perrett DI, Johnston RS, Benson PJ, Jordan TR, Heeley DW, Bettucci D, Mortara F, Mutani R, Terazzi E, Davidson DLW (1991) Perception and action in "visual form agnosia". Brain 114:405–428

Milner AD, Paulignan Y, Dijkerman HC, Michel F, Jeannerod M (1999) A paradoxical improvement of misreaching in optic ataxia: new evidence for two separate neural systems for visual localization. Proc R Soc Lond B Biol Sci 266:2225–2229

Namba J, Baldo VC (2004) The modulation of the flash-lag effect by voluntary attention. Perception 33:621–631

Nijhawan R (1994) Motion extrapolation in catching. Nature 370:256–257

Nijhawan R, Kirschfeld K (2003) Analogous mechanisms compensate for neural delays in the sensory and the motor pathways. Evidence from motor flash-lag. Curr Biol 13:749–753

Nishida S, Johnston A (1999) Influence of motion signals on the perceived position of spatial pattern. Nature 397:610–612

Rizzolatti G, Matelli M (2003) Two different streams form the dorsal visual system: anatomy and functions. Exp Brain Res 153:146–157

Rossetti Y, Pisella L, Vighetto A (2003) Optic ataxia revisited: visually guided action versus immediate visuomotor control. Exp Brain Res 153:171–179

Snowden RJ (1998) Shifts in perceived position following adaptation to visual motion. Curr Biol 8:1343–1345

Ungerleider LG, Mishkin M (1982) Two cortical visual systems. In: Ingle DJ, Goodale MA, Mansfield RJW (Eds) Analysis of visual behavior. MIT Press, Cambridge MA

Weiskranz L (1986) Blindsight: a case study and its implications. Oxford Press, Oxford

Whitney D (2005) Motion distorts perceived position without awareness of motion. Curr Biol 15:R324–326

Whitney D, Cavanagh P (2000) Motion distorts visual space: shifting the perceived position of remote stationary objects. Nat Neurosci 3:954–959

Whitney D, Murakami I (1998) Latency difference, not spatial extrapolation. Nat Neurosci 1:656–657

Whitney D, Murakami I, Cavanagh P (2000) Illusory spatial offset of a flash relative to a moving stimulus is caused by differential latencies for moving and flashed stimuli. Vision Res 40:137–149

Whitney D, Goltz HC, Thomas CG, Gati JS, Menon RS, Goodale MA (2003a) Flexible retinotopy: motion-dependent position coding in the visual cortex. Science 302:878–881

Whitney D, Westwood DA, Goodale MA (2003b) The influence of visual motion on fast reaching movements to a stationary object. Nature 423:869–873

Yamagishi N, Anderson SJ, Ashida H (2001) Evidence for dissociation between the perceptual and visuomotor systems in humans. Proc R Soc Lond B Biol Sci 268:973–977

14
Neural Substrates of Action Imitation Studied by fMRI

Shigeki Tanaka

1 Introduction

Action imitation presents several interesting points for the study of object recognition and action because, during imitation, a person must manipulate body parts as objects and, at the same time, do this using our own body. Other people's action can be imitated so easily that the complicated cognitive processes involved in imitation tend to be overlooked. It is well known that even the neonate can imitate (Meltzoff and Borton 1979; Meltzoff and Moore 1983). The process of imitating other people's actions involves various cognitive elements such as visual perception of target actions, transforming perceived actions into one's own body and/or motor representation, and simulation of one's own motor image (Decety and Chaminade 2003; Jackson and Decety 2004; Chaminade et al. 2005).

Mirror neurons were reported to be activated when a monkey observed the hand actions of another as well as when the monkey himself performs the same action (Rizzolatti et al. 1996), implying that these neurons represent the concept of action regardless of who performs the action. The impact of the finding of mirror neurons was very strong, and its concept has been introduced into various topics in cognitive science from motor cognition to the problems of self and others (Rizzolatti and Craighero 2004), such as a sense of action agents (Decety et al. 2002) and empathy for others (Carr et al. 2003; Gallese 2003).

In this chapter, two fMRI studies concerning action imitation (Tanaka et al. 2001; Tanaka and Inui 2002) and recent related findings by other groups are introduced.

Department of Psychology, Faculty of Human Sciences, Jin-Ai University, 3-1-1 Ode-cho, Echizen, Fukui 915-8586, Japan

2 Experiment 1: Finger Action Imitation with and without Symbolic Meaning

2.1 Ideomotor Apraxia

A pathological state exhibiting deficits in pantomiming on verbal command or gestural imitation is called ideomotor apraxia (IMA) and is caused by left inferior parietal lesions. Heilman et al. (1982) proposed that the motor engrams for skilled movements are stored in the left supramarginal gyrus. Then damage to this area or disconnection between this area and the motor area lead to deficits of action imitation. Recently, Goldenberg and Hagmann (1997) showed that IMA was caused by poor perception of target postures. In their study, they examined whether IMA patients could manipulate mannequins by instructing the patients to imitate presented actions using mannequins. The action required the patients to properly imitate a target action different from those required to manipulate mannequins. The IMA patients were poor at this task as well as at tasks that required imitation using their own bodies. This means that the poor imitation by IMA patients cannot be attributed to the damage to the stored motor engrams, but might be derived from a deficit in perceiving target actions.

2.2 fMRI Study of Finger Action Imitation

Previous neuroimaging studies investigating on the imitation of finger actions used very simple tasks; subjects were required to raise their fingertips slightly according to the presentation of line-drawn or photos of finger pictures (Iacoboni et al. 1999; Krams et al. 1998). In those studies various brain activations were reported, including that in the parietal area. However, the stimuli were too simple to study the elements of the imitation process, such as the detailed analysis and perception of target posture, transformation from perceived posture into one's own body image or manipulation of the motor image to produce real action. In order to study those elements in action imitation, we performed an fMRI study using rather complicated tasks, i.e., finger configurations with or without symbolic meaning. During imitating of actions with symbolic meanings, visual recognition might lead to activate stored motor memories of those actions. However, imitating novel actions requires more detailed observation to understand the spatial relations among fingers and also requires more detailed motor control for action execution.

2.2.1 Tasks and Subjects

An imitation task with three conditions was used. In the first (S–) condition, 10 pictures of meaningless (in Japanese culture) finger configurations were presented to the subjects (three such items are shown in Fig. 1a). The subjects were required to imitate the finger configuration using their right hand, during the

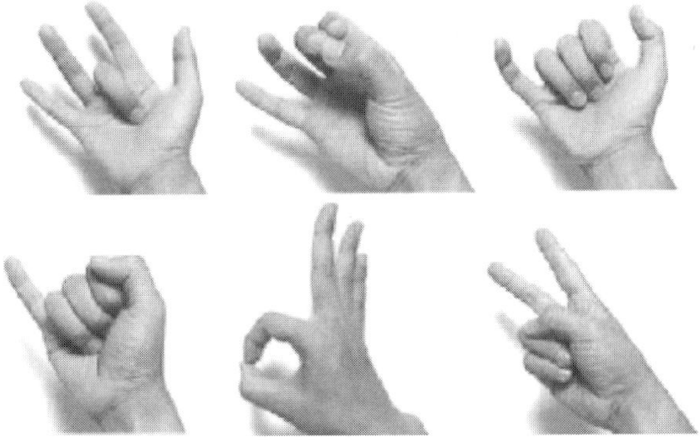

a

b

FIG. 1. Examples of visual stimuli. **a** Finger configuration without symbolic meaning (S–, top row), **b** With symbolic meaning (S+, bottom row). Three of the ten pictures are shown for each condition. In **b**, the usual meanings of the configurations are promise, OK, and scissors from left to right

stimulus presentation. Stimuli were presented for 2 seconds for each (SOA = 3 s, ISI = 1 s, 10 pictures in random order per block, block duration = 30 s). The second (S+) condition, 10 pictures of finger configurations (Fig. 1b) with symbolic meaning (in Japanese culture), were presented in a manner identical with that for the S– condition. The third condition was a rest condition: a fixation point was shown instead of finger pictures with the same SOA and ISI. Subjects were instructed just to watch the fixation point. The three conditions were repeated four times in a counter-balanced order. The visual stimuli were controlled by a personal computer and were projected onto a screen by a liquid crystal display projector seen through a mirror set above their eyes as the subjects lay in the MRI machine. The visual angle was 5.3 × 5.4. The subjects' performance was monitored through the window at the MRI control console. All responses were evaluated as correct or incorrect and recorded into the list of the stimuli for each subject. Responses were evaluated as correct whenever fingers to be extended and those to be folded were correctly imitated.

A total of nine right-handed subjects (six male and three female; mean age 25.2 years; range 22–34) participated. All subjects were fit, healthy, on no medication, free from any history of neurological or psychiatric illness and gave written informed consent.

2.2.2 fMRI

A 1.5 T MRI scanner was used to acquire 72 scans per subject with a gradient echo EPI sequence (TR/TE = 5000/40 ms, FA = 90, FOV = 220 mm, matrix = 64 × 64, 32 axial slices, 5 mm slice thickness without gap). The first four scans were discarded to avoid initial instability. Data analysis was performed using SPM 96 (http://www.fil.ion.ucl.ac.uk). All EPI images were spatially normalized with MNI template for group analysis. Imaging data were corrected for head movements and signal intensity variation and smoothed with an isotropic Gaussian kernel 10 mm FWHM. Significance was assessed using the delayed box-car reference.

2.3 Results and Discussion

The mean number of incorrect responses was 3.4 (range 2–5) among 80 responses. All incorrect responses were made in the S– condition and most of them were within the first S– block. Most of the incorrect responses were a timeout type (subjects failed to make any response or made uncertain finger movement, then skipped to the next image). This result was consistent with the subjects' comments after the experiment that an effort was required to imitate meaningless finger configurations without seeing their own hand, even though it had been confirmed that they could imitate all stimuli in both the S– and S+ conditions promptly and completely in front of a PC monitor outside the MRI room.

For fMRI data analysis, In S– vs. rest, there was activation in the right SMG (Fig. 2a) which was not observed in S+ vs. rest. The comparison between S– and S+ is shown in Fig. 2b (uncorrected for multiple comparisons). In the comparison S– vs. S+, only bilateral parietal activation differed significantly. There was no significant difference of activation detected in the comparison of S+ vs. S–. In both comparisons of S– vs. rest and S+ vs. rest, strong right cerebellar activation was observed.

In the comparison of S– vs. S+ conditions, significant activation in the bilateral SMG was observed. In comparison with the rest condition, both S– and S+ conditions showed activation in the left SMG, but only the S– condition showed activation in the right SMG. According to interviews after the experiment, subjects did

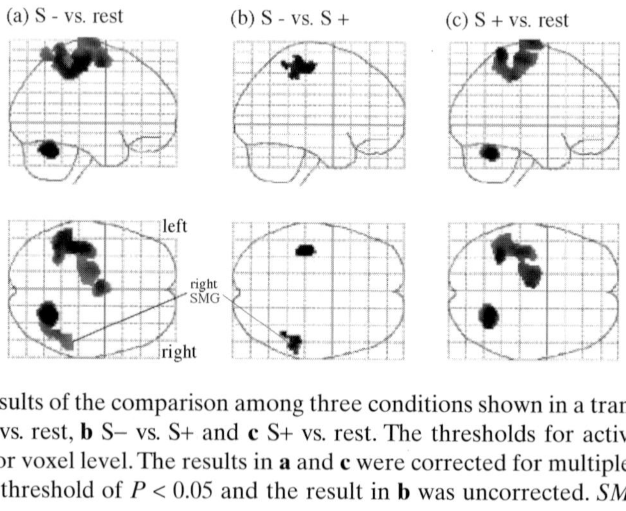

Fig. 2. The results of the comparison among three conditions shown in a transparent brain system. **a** S– vs. rest, **b** S– vs. S+ and **c** S+ vs. rest. The thresholds for activation was set at $P < 0.001$ for voxel level. The results in **a** and **c** were corrected for multiple comparisons at the extent threshold of $P < 0.05$ and the result in **b** was uncorrected. *SMG*: supramarginal gyrus

not have to carefully analyze the position of each finger in the S+ condition, because they were well accustomed to the stimuli. Contrarily the S– condition required more detailed visual analysis of the target stimuli as well as somesthetic analysis/integration than the S+ condition. Iacoboni et al. (1999) also suspected that activation in the right SMG in their fMRI study of finger imitation might imply that the perceived information of the observed action, such as the angle of a finger joint, is stored in this area. We suppose the activation in the right supramarginal gyrus (SMG) detected in S– vs. S+ might be related to the perception of target actions of the S– condition that required more detailed visual analysis than under the S+ condition. The subjects had to feel their own fingers because they could not see them during the tasks, especially imitating unaccustomed finger configurations under the S– condition. The difference in activation in the left primary motor and sensory areas on comparison of S– vs. S+ might reflect this effort. Sensory and/or motor representation of fingers might be necessary for the manipulation of mental representations required under the S– condition. The left SMG activation in the comparison S– vs. S+ might show the deep involvement of this area in preparing and executing novel finger configurations which require integrating several simple actions (each finger posture) into a more complex one. Lesions in the inferior parietal lobule are known to cause disturbances in complex polymodal integration of somesthetic and visual representation such as ideomotor apraxia (Heilman et al. 1982; Ochipa et al. 1994; Goldenberg et al. 1996). Our result is consistent with these neuropsychological findings.

3 Hand/Arm Action vs. Finger Configuration

3.1 fMRI Study of Action Imitation: Hand/Arm vs. *Finger Action*

Goldenberg (1999) reported an interesting clinical study that patients with right brain damage performed more poorly in finger configuration imitation than in hand/arm action imitation, and vice versa for patients with left brain damage. A neuroimaging study showed different parietal involvement in the recognition tasks of hand/arm action and finger configuration (16) (Hermsdorfer et al. 2001). To study the neural substrates involved in action imitation of hand/arm and those of finger configurations, we performed an fMRI study.

3.1.1 Subjects and Task

A total of 12 right-handed subjects (six male; mean age = 24.8, range 21–34 years old) participated in a task with three conditions. In all conditions, subjects were instructed to imitate presented postures using their right hand or fingers. The first condition was a control condition (rest) in which pictures such as shown in Figure 3a were presented for 2 sec followed by a fixation point for 1 sec, which

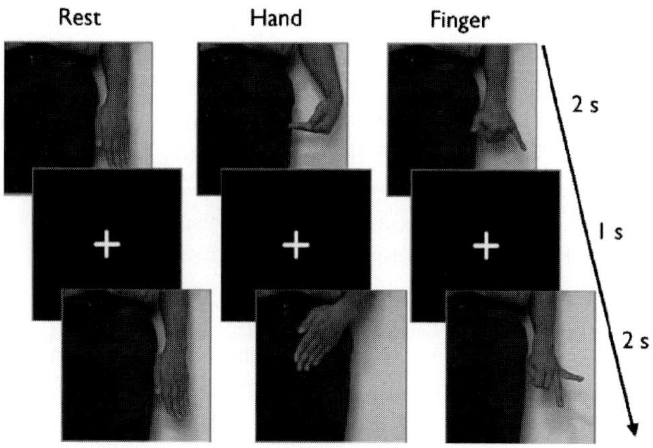

FIG. 3. The experimental design for the three conditions

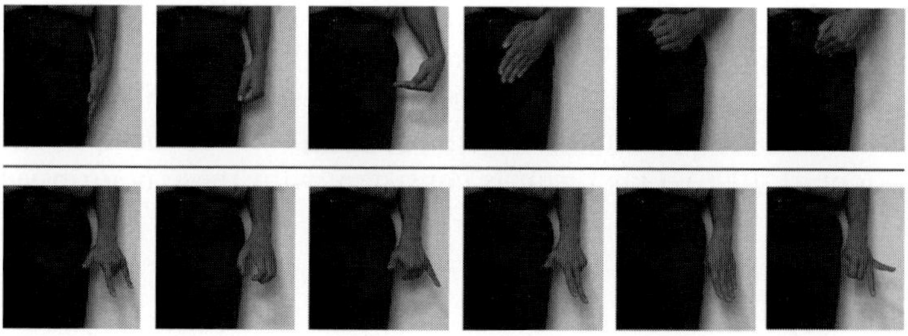

FIG. 4. Visual stimuli. Hand/arm postures (top row) and Finger configuration (bottom row)

was repeated 10 times in one block. Subjects were instructed just to watch it without any motion. In the second (hand) condition, one of six pictures of meaningless hand/arm postures was presented to the subjects (Fig. 4, top row). There were two patterns of elbow joint angle, straight or bent, and three patterns of hand position. Subjects were required to imitate the presented posture using their right arm and hand; stimulus movements were performed with the left hand of the demonstrator, so that subjects imitated the movement as if they were seeing themselves in a mirror. At each trial, as the stimulus disappeared, they were required to go back to the rest position. Stimuli were presented for 2 s each (SOA = 3 s, ISI = 1 s, 10 pictures in random order per block, block duration = 30 s). In the third (finger) condition, one of six pictures of meaningless finger configurations (Fig. 4, bottom row) was presented in the same fashion as in the hand condition. The hand and the finger conditions were repeated four times in a counter-balanced order. The presentation and the control system of visual stimuli were the same as used in the experiment 1. Subjects' responses through all sessions were recorded by a digital video camera for estimating their performance.

3.1.2 fMRI

A 1.5 T MRI scanner was used. A total of 100 scans per subject were acquired with a gradient echo EPI sequence (TR/TE 5000/55 ms, FA = 90, FOV 240 mm, matrix 64 × 64, 38 axial slices, 5 mm slice thickness without gap). The first four scans were discarded to avoid initial instability. Data analysis was performed using SPM 99 (http://www.fil.ion.ucl.ac.uk). All EPI images were acquisition-corrected for sampling bias effects caused by different times relative to the haemodynamic response for each subject. The images were realigned to correct for interscan movement and spatially normalized with MNI template for group analysis. The images were smoothed with an isotropic Gaussian kernel of 8 mm FWHM. Significance was assessed using the delayed box-car reference convolved with a haemodynamic response function. Linear contrasts between different conditions gave results as activated areas by creating a spatially distributed map of the t-statistic (SPM{t}). Activation was thresholded at $P < 0.001$, corrected for multiple comparisons for each subject. The acquired four contrast maps, hand vs. rest, finger vs. rest, hand vs. finger and finger vs. hand, of each subject were jointly used for group analysis based on the random effects analysis (17) (Friston et al. 1999).

3.2 Results and Discussion

All subjects' responses recorded by the digital video camera were evaluated by a naive observer and it was determined that all subjects responded correctly on all trials. Compared with the control condition, in the hand condition a significant activation was detected in the bilateral precentral and postcentral gyri, inferior parietal gyri, and cerebellum, and in the right occipital lobe, thalamus and putamen. Under the finger condition, significant activation was detected in various areas including the bilateral pre/postcentral gyri and the inferior frontal gyrus (Brodmann area (BA) 44).

When the hand condition was compared with the finger condition, significant activation was detected in the bilateral superior parietal lobule and the bilateral pre/post central gyri (Fig. 5, left). The finger condition compared with the hand condition demonstrated significant activation in the left inferior frontal area (Brodmann area (BA) 44, 47), bilateral inferior parietal lobules and right superior parietal lobule (Fig. 5, right).

3.2.1 Broca's Area

Significant activation was observed in Broca's area under the finger condition compared with the rest condition or with the hand condition, suggesting that Broca's area might be more important in imitating finger actions than in that of hand/arm actions. Since the first report of mirror neurons by Rizzolatti et al.

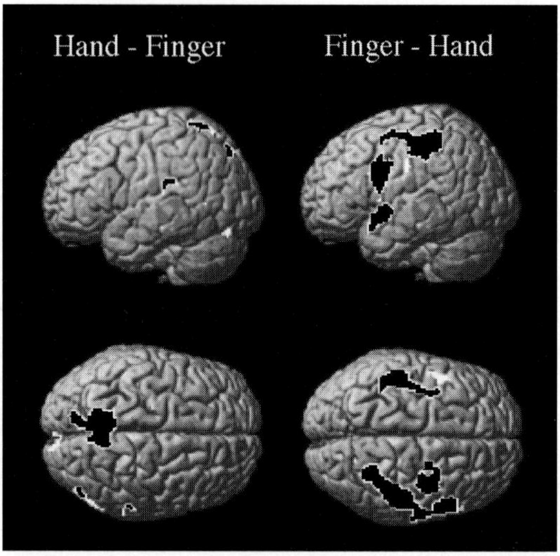

FIG. 5. The results of the random effects analysis on 12 subjects shown on a template brain image. Hand vs. finger (left) and finger vs. hand (right) are shown. The statistical threshold is voxel level $P < 0.001$ (uncorrected)

(1996), many studies have indicated involvement of Broca's area in human action imitation or observation (Iacoboni et al. 1999; Iacoboni et al. 2001; Krams et al. 1998; Perani et al. 2001, etc.). Broca's area activation was not reported in some imaging studies in which task conditions include recognition of hand/finger postures or simple observation of another person's grasping action (Hermsdorfer et al. 2001; Perani et al. 2001). Buccino et al. (2001) reported in their fMRI study that Broca's area was activated when subjects observed hand actions that actually manipulated objects but not when subjects observed pantomimes of object manipulation. In the PET study by Hermsdorfer et al. (2001), neither task conditions of recognizing hand/arm action vs. recognizing finger configuration showed any Broca's area activation. In our study, subjects actually performed action imitation during the scan and Broca's area activation was detected under the finger condition. Consequently, Broca's area might be involved more in the process of execution than in the process of recognition in human action imitation.

3.2.2 Parietal Lobe

Goldenberg (1999) reported that the laterality of brain lesion showed a correlation with action imitation tasks of hand/arm action and finger configuration. He proposed that imitations of hand and finger gestures are subserved by at least partially different mechanisms that are differently distributed across the two hemispheres: the right hemisphere being more involved in the process of visuospatial cognition of presented gestures, while the left hemisphere is more involved in the process of referring to knowledge of one's own body as well as in the process of preparing and executing one's own action. In their PET study of the recognition of other people's actions, Hermsdorfer et al. reported that the recognition of finger configurations showed more symmetrical activation in the parietal area, while that of hand postures showed left lateralized parietal activation, which is consistent with their clinical observations (Hermsdorfer et al. 2001). In the present study, the activated area in the parietal lobe was located mainly in the left hemisphere under the hand condition and bilaterally under the finger condition. These findings are consistent with the clinical findings reported by Goldenberg (1999). The right parietal area was more involved in finger action imitation which might require greater recognition of the spatial relation of the presented fingers. Choi et al. (2001) reported left superior parietal activation in their fMRI study in which subjects pantomimed tool use. The activation pattern shown in their result were similar to that under the hand condition of our study, possibly due to the fact that most of the actions using tools consist of not finger but hand/arm actions.

Why are there such differences between the cortical networks involved in hand/arm posture and those of finger configuration imitation? We suspect it might be due to differences in the modality of feedback information (visual vs. somatosensory) during development. That is, one can see one's own fingers during imitation of another person's finger configuration, but one cannot see one' s own body movement during the imitation of hand/arm postures without using a mirror. Thus visual feedback is more important for finger action imitation, while somatosensory feedback plays the main role in imitating hand/arm posture. It may therefore be that through the developmental process, different cortical networks come to be involved in those two types of action imitation. Goldenberg (2002) reported recently that a patient group with right brain damage performed more poorly in foot gesture imitation than in hand gesture imitation. This result can also be explained by the availability of visual feedback during imitation; it is easy to see one's own foot gesture, but it is difficult to see one's own hand/arm gestures (and their spatial relation to one's own face) by oneself.

With regard to the coordinates of mental representation, hand/arm posture is related more to self-centered coordinates; the arms and hands might be represented using spatial relations referred to the body. Contrarily, object-centered coordinates might be more important for the representation of finger configurations, the mutual spatial relations among the fingers being more important than their relation to the body.

Visuoconstructive disturbance caused by right hemisphere damage is a defect in copying geometric figures. Visuoconstructive functions consist of the cognitive process of recognizing the spatial relations among objects and expressing these relations as one's motor output. It has been reported that patients with visuoconstrutive disturbance showed deficits in imitating finger configurations (Yamadori 1985; Ogura and Yamadori 1983). This clinical finding might also imply that one important element of finger action imitation is recognizing spatial relations among presented objects.

References

Buccino G, Binkofski F, Fink GR, Fadiga L, Fogassi L, Gallese V, Seitz RJ, Zilles K, Rizzolatti G, Freund HJ (2001) Action observation activates premotor and parietal areas in a somatotopic manner: an fMRI study. Eur J Neurosci 13:400–404

Carr L, Iacoboni M, Dubeau MC, Mazziotta JC, Lenzi GL (2003) Neural mechanisms of empathy in humans: a relay from neural systems for imitation to limbic areas. Proc Natl Acad Sci U S A 100:5497–5502

Chaminade T, Meltzoff AN, Decety J (2005) An fMRI study of imitation: action representation and body schema. Neuropsychologia 43:115–127

Choi SH, Na DL, Kang E, Lee KM, Lee SW, Na DG (2001) Functional magnetic resonance imaging during pantomiming tool-use gestures. Exp Brain Res 139:311–317

Decety J, Chaminade T (2003) When the self represents the other: a new cognitive neuroscience view on psychological identification. Conscious Cogn 12:577–596

Decety J, Chaminade T, Grèzes J, Meltzoff AN (2002) A PET exploration of the neural mechanisms involved in reciprocal imitation. NeuroImage 15:265–272

Friston KJ, Holmes AP, Worsley KJ (1999) How many subjects constitute a study? Neuroimage 10:1–5

Gallese V (2003) The roots of empathy: the shared manifold hypothesis and the neural basis of intersubjectivity. Psychopathology 36:171–180

Goldenberg G (1999) Matching and imitation of hand and finger postures in patients with damage in the left or right hemispheres. Neuropsychologia 37:559–566

Goldenberg G, Hagmann S (1997) The meaning of meaningless gestures: a study of visuo-imitative apraxia. Neuropsychologia 35:333–341

Goldenberg G, Hermsdorfer J, Spatt J (1996) Ideomotor apraxia and cerebral dominance for motor control. Brain Res Cogn Brain Res 3:95–100

Goldenberg G, Strauss S (2002) Hemisphere asymmetries for imitation of nevel gestures. Neurology 59:893–897

Heilman KM, Rothi LJ, Valenstein E (1982) Two forms of ideomotor apraxia. Neurology 32:342–346

Hermsdorfer J, Goldenberg G, Wachsmuth C, Conrad B, Ceballos-Baumann AO, Bartenstein P, Schwaiger M, Boecker H (2001) Cortical correlates of gesture processing: clues to the cerebral mechanisms underlying apraxia during the imitation of meaningless gestures. Neuroimage 14:149–161

Iacoboni M, Woods RP, Brass M, Bekkering H, Mazziotta JC, Rizzolatti G (1999) Cortical mechanisms of human imitation. Science 286:2526–2528

Iacoboni M, Koski LM, Brass M, Bekkering H, Woods RP, Dubeau MC, Mazziotta JC, Rizzolatti G (2001) Reafferent copies of imitated actions in the right superior temporal cortex. Proc Natl Acad Sci U S A 98:13995–13999

Jackson PL, Decety J (2004) Motor cognition: a new paradigm to study self-other interactions. Curr Opin Neurobiol 14:259–263

Krams M, Rushworth MF, Deiber MP, Frackowiak RS, Passingham RE (1998) The preparation execution and suppression of copied movements in the human brain. Exp Brain Res 120:386–398

Meltzoff AN, Borton RW (1979) Intermodal matching by human neonates. Nature 282:403–404

Meltzoff AN, Moore MK (1983) Newborn infants imitate adult facial gestures. Child Dev 54:702–709

Ochipa C, Rothi LJ, Heilman KM (1994) Conduction apraxia. J Neurol Neurosurg Psychiatry 57:1241–1244
Ogura J, Yamadori A (1983) Finger imitation difficulty constructional disorder and classical apraxias. No To Shinkei 35:759–763
Perani D, Fazio F, Borghese NA, Tettamanti M, Ferrari S, Decety J, Gilardi MC (2001) Different brain correlates for watching real and virtual hand actions. NeuroImage 14:749–758
Rizzolatti G, Craighero L (2004) The mirror-neuron system. Annu Rev Neurosci 27:169–192
Rizzolatti G, Fadiga L, Gallese V, Fogassi L (1996) Premotor cortex and the recognition of motor actions. Brain Res Cogn Brain Res 3:131–141
Tanaka S, Inui T (2002) Cortical involvement for action imitation of hand/arm postures versus finger configurations: an fMRI study. Neuroreport 13:1599–1602
Tanaka S, Inui T, Iwaki S, Konishi J, Nakai T (2001) Neural substrates involved in imitating finger configurations: an fMRI study. Neuroreport 12:1171–1174
Yamadori A (1985) Introduction to Neuropsychology (in Japanese). Igakushoin, Tokyo, p 55

15 Two Types of Anticipatory-Timing Mechanisms in Synchronization Tapping

YOSHIHIRO MIYAKE[1], YOHEI ONISHI[1], and ERNST PÖPPEL[2]

1 Introduction

Mutual coordination of timing is required to produce synchronous cooperative behavior between humans, and an anticipation mechanism related to external events is thought to be indispensable to generate such movement. The importance of this timing control becomes clear if one considers, for example, playing together in a musical ensemble. However, it has been reported that a time difference exists between awareness of cognitive synchrony and physical synchrony, such as a negative asynchrony phenomenon (see next paragraph). Analysis of this anticipatory mechanism should be performed, not only to elucidate the physical process, but also to understand the underlying cognitive process in which a higher brain function, such as attention (Kahnemann 1973), is involved.

The synchronization tapping task has been used as the simplest method for examining the timing mechanism. In this experiment, the subject is required to synchronize his/her finger movement with a periodic auditory or visual stimulus. The most striking demonstration of anticipatory timing control occurs when the onset of each tap precedes the onset of stimulus by several 10ms (Stevens 1886; Woodrow 1932; Fraisse 1966; Kolers and Brewster 1985; Peters 1989; Mates et al. 1994; Aschersleben and Prinz 1995). This pressing-in-advance phenomenon, of which the subject is unaware, demonstrates that the motor command to the finger is generated before the onset of the auditory stimulus, suggesting a process of anticipatory timing control. The negative time offset caused by tapping in advance is referred to as negative asynchrony – a phenomenon that is always observed in the synchronization tapping task in response to a periodic stimulus.

To examine this type of phenomenon, Mates et al. (1994) conducted a synchronous tapping experiment using a periodic auditory stimulus within a range of 300 to 4,800ms. They confirmed that negative asynchrony was observed for

[1]Department of Computational Intelligence and Systems Science, Tokyo Institute of Technology, Midori, Yokohama 226-8502, Japan
[2]Institute of Medical Psychology, Ludwig-Maximilian University of Munich, Goethestr. 31, Munich 80336, Germany

all of the above stimulus intervals with a difference in the degree of its occurrence. They found that the upper limit for the generation of stable, negative asynchrony with little fluctuation is 2 to 3 s for the interstimulus-onset interval (ISI). It was also reported that if the ISI limit is exceeded, reactive responses become mixed with the negative asynchrony.

Synchronization tapping tasks and other types of time-discrimination tasks and time-reproduction tasks (Ivry 1996, 1997; Pascual-Leone 2001; Rao et al. 1997) have demonstrated that the cerebellum plays an important role in neural mechanisms that support perception of time intervals under 1 s. Higher brain functions contribute to the perception of time intervals that exceed 2 to 3 s (Kagerer et al. 2002; Brown 1997). Mangles et al. (1998) conducted a series of experiments on time perception under 2 sets of conditions – short (400 ms) and long (4 s) time intervals – in subjects with injuries to the cerebellum and prefrontal cortex. They found that subjects with an injury to the prefrontal cortex exhibited a deterioration of performance only on the long-duration discrimination tasks. They also discovered a deficiency in the subjects' working memory. These findings suggest a multi-component timing mechanism (Ivry 1997) and the importance of working memory in the perception of long time periods.

Experiments by Mates et al. (1994) did not clarify the role of working memory or the contribution of these two types of timing mechanisms to the occurrence of negative asynchrony. Miyake et al. (2001) proposed the hypothesis of a dual-anticipation mechanism in sensory-motor coupling. An experiment supporting this hypothesis was recently reported by Zelaznik et al. (2002). The experiment presented here was designed to determine the effects of higher brain functions like attention on a synchronization tapping task.

A number of cognitive models have been proposed to explain the relationship between the perception of a time interval exceeding 2 to 3 s and attention. Among these, the "attention-allocation model" is based on the premise that decision-making time is determined by the extent of attentional resources allocated to the temporal-information processing system in contrast to the mental-activity processing system unrelated to time (nontemporal information processing) (Brown 1997; Macar and Casini 1999). Central activation of working memory is involved in this allocation of attention (Baddeley 1986, 1998a, b; Osaka 2000). According to Kahnemann's attention-capacity model (1973), there are limited attentional resources, and these resources determine the limits in the processing of perceptual information. Attention is a critical resource in the execution of mental activities, and it can be appropriately allocated to each separate task according to the tendencies and intentions of each individual during the simultaneous execution of multiple tasks. In this condition, it is possible to quantify the amount of the attentional resources that has been allocated based on the magnitude of the mental processing involved.

We examined the range of ISI affected by attention in a synchronization tapping task based on the above models. If the subject's attention is directed toward processing of information other than tapping during a synchronization tapping task, it becomes difficult for the subject to focus the amount of attention required for the execution of the tapping task due to the limited capacity of attentional resources. If the amount of attention required in the tapping task exceeds the remaining resources, sufficient processing resources cannot be allocated to the temporal-information processing system, the ability to make temporal decisions becomes disrupted, and anticipatory timing control is thought to be affected.

2 Methods

A dual-task method (Baddeley 1986) was used to control the subject's attention. In this experiment, the processing capacity required for executing the primary task was reduced by having the subject engage in an additional (or secondary) task while still performing the primary task. Well-known examples of these types of test are the reading-span test (Daneman and Carpenter 1980; Osaka and Osaka 1994), which measures the capacity of working memory when a subject is simultaneously reading a short sentence aloud and engaged in a word-memory task. Another is the articulatory-suppression method, which examines the organization of coding of auditory information when a subject is engaged in a cognitive activity like memory while simultaneously repeating a word, such as "a" or "the" (Saitoh 1997). We employed a word-memory task as the secondary task to control the subject's attention.

The word-memory task was used to restrict the target of attention control to short-term memory and to determine the correlation between attention and negative asynchrony in the synchronization tapping task. This type of transient memory has been regarded as a function of working memory and is often employed as a secondary task to divert the attentional resources of the subject. In this study, the difference in the number of memorized words was regarded as the difference in the amount of attentional resources and attention capacity that was available in the tapping task. The memory task involved two different numbers of words as a secondary task. If the attention capacity required by the memory task corresponds to the processing resources that are used in the synchronization tapping task, some type of interference would appear between the two, and the difference in the number of memorized words is thought to reflect the occurrence rate of negative asynchrony.

The subjects were asked to press a button in synchrony with the onset of a periodic pulse auditory stimulus as their primary task. A total of ten different ISIs were used in this study, and this task was performed under the following two conditions. Each trial had a fixed ISI auditory stimulus for the controlled condition (N condition), then repeated for each of the ten durations of ISI. During the trials, the subjects were required to manually press a button precisely at stimulus onset. The word-memory task (M condition) was conducted parallel to the the control task (N condition). The details are explained in the following section(s).

2.1 Tapping Task

The subjects were all right-handed and were required to press a button with their right index finger in synchrony with the onset of a periodic pulse auditory stimulus. A total of ten different ISIs were used in this study: 450, 600, 900, 1,200, 1,500, 1,800, 2,400, 3,600, 4,800, and 6,000 ms. The sequence of ISIs was randomized for each subject. The duration of each auditory stimulus was 100 ms, and the frequency was 500 Hz. The acoustic pressure was set at an appropriate magnitude that allowed the subjects to clearly hear the auditory stimulus. It was the same for each subject throughout all the trials.

2.2 Definition of Parameters

The data measured during this experiment were stimulus onset and tap onset. The main target of analysis was the time difference between the stimulus onset and the tap onset, defined as synchronization error (SE). This reflects the temporal relationship between stimulus and action. A positive SE indicates that the tapping onset lagged behind the stimulus onset. As demonstrated by Mates et al. (1994), tapping can be divided into 2 types, that with negative asynchrony and that reactive to stimulus. Therefore, the former is referred to as anticipatory tapping and the latter, as reactive tapping. The relationship between these two parameters is shown in the Figure 1a.

2.3 Subjects

Six healthy male university graduate students in their 20s volunteered to participate in this study. They all had experience in synchronization tapping tasks. All of the subjects were right-handed and had normal hearing.

2.4 System

The system used in this experiment was loaded onto a personal computer with a single task OS (PC-DOS2000, IBM). The stimulus sound was transmitted to the subjects via headphones from an external sound source connected to the PC through a parallel port. In addition, the button that the subjects pressed was connected to the PC via a parallel port. The program used in the study was developed using the programming language C. A built-in real time clock (RTC) with a time resolution of 1 ms was used to measure the time when the button was pressed and the time of auditory stimulus presentation.

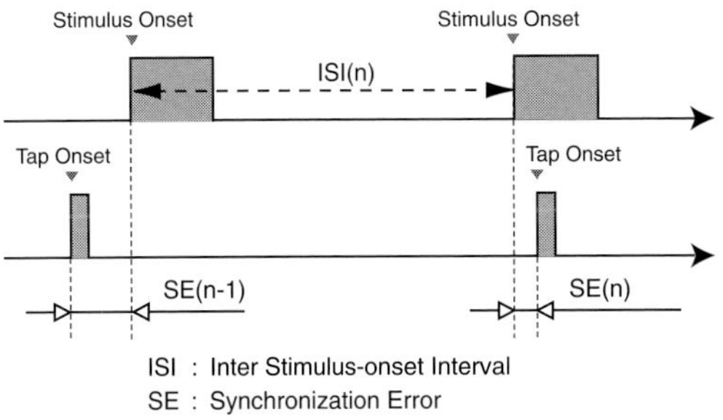

a

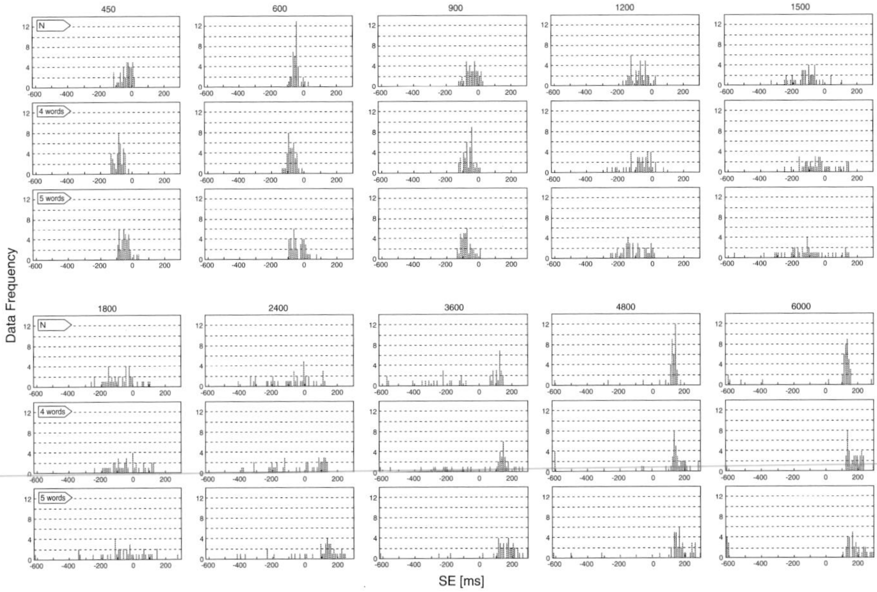

b

FIG. 1. Synchronization Error distribution. **a** Temporal relationship between tapping onset and stimulus onset. The time difference between the stimulus onset and the tap onset was defined as Synchronization Error *(SE)*. Negative SE indicates that tapping precedes the stimulus onset and corresponds to anticipatory tapping. The time difference between two successive stimulus onsets was defined as the Interstimulus-onset Interval *(ISI)*. The duration of each stimulus was 100 ms. **b** SE distribution for every Interstimulus-onset interval *(ISI)* of subject A is shown. The upper figure corresponds to the normal condition, the lower two figures correspond to the memory condition. Here N represents normal synchronization tapping, and 4 words or 5 words represent tapping with 4- or 5-word memory tasks, respectively. The number at the head of each figure represents ISI [ms]. (From Miyake et al. 2004, with permission)

2.5 Procedure

The task was to press a button in coordination with a periodic-pulse auditory stimulus. This task was conducted under the following two conditions:

(1) N (control) condition: Each trial consisted of a set ISI auditory stimuli and was conducted for ten different ISIs. During each trial, the subjects were requested to press a button the moment they heard an auditory stimulus. Each trial lasted 1 minute so that a memory task could be performed as a secondary task. By changing the number of trials corresponding to the ISIs, data from a total of 40 taps could be collected for each ISI. Since the objective was to observe a steady reaction in the subjects, data recording began 10s after the onset of the initial tap in each trial.

(2) M (memory-task) condition: Tapping was performed in the same manner as under the N condition in parallel with the word-memory task. The subjects were asked to remember a word using a Japanese phonetic character, which consisted of 3 to 5 morae. A "morae" is a syllable representing a Japanese word. All of the words were meaningful, but the combinations used in each trial were selected to make it difficult to create meaningful associations between words. In addition, the subjects were instructed not to memorize the words using the storytelling method (a method of memorization in which a story is created using the displayed words to shift the words into long-term memory). Either four or five words were displayed in each trial. The mean number of morae was 3.69 for the 4-word condition and 3.68 for the 5-word condition. The trials commenced simultaneously when the subject pressed the space bar on the computer keyboard. Once the space bar was pressed, the word set was displayed in the center of the monitor screen (IBM ThinkPad 535) for 3s. The monitor then blacked out, and an auditory stimulus was immediately presented. The subjects were required to perform tapping for a 1-minute period while remembering the words. Immediately after completion of the tapping, the subjects were asked to recite the retained words. The order of the words was not considered relevant. Subjects A, B, and C performed the experiment in the order of the N condition – 4-word condition followed by 5-word condition, whereas subjects D, E and F performed the experiment in the order of the N condition – 5-word condition followed by 4-word condition.

The subjects were also instructed not to time the tapping by counting to themselves while tapping or by making rhythmic physical movements. Each trial was conducted after a suitable interval to ensure that the subject's concentration was not adversely affected by fatigue resulting from the preceding trials.

3 Results

3.1 Correct Response Rate for Word-Memory Task

The correct response rates for the word-memory tasks for each subject are shown in Table 1. The values for each subject are the mean values for each trial. The

TABLE 1. Correct response rates for memory task. The value for each subject is the subject's average value of all trials (from Miyake et al. (2004), with permission)

	Correct response rate for memory task	
Subject	4 words (%)	5 words (%)
A	100.0	96.4
B	92.0	77.3
C	98.9	90.9
D	100.0	94.6
E	98.9	92.8
F	100.0	98.2
Average	98.3	91.7

correct response rate among subjects was 98.3% for 4 words and 91.7% for 5 words. The difference between the mean values for the 2 groups was significant at $P < 0.05$ on Wilcoxon sign rank sum test. There was an exceptionally large drop in performance observed for subject B. Memorization of 4 words could be executed almost perfectly by each of the subjects, whereas there was a difference in scores for the 5-word memorization task, which appeared to be more difficult. This result suggests that the attentional resources required to memorize 5 words exceeded or was close to the capacity limit.

3.2 Distribution of Synchronization Errors (SE)

The data obtained in this experiment were stimulus onset and tap onset. Synchronization error (SE), the time difference between the stimulus onset and the tap onset, was analyzed as an index reflecting the temporal relationship between stimulus and response.

The SE distribution at each ISI is shown in Figure 1b for Subject D. The negative SE indicates that the tap precedes the auditory stimulus. The shape of the SE distribution for the N condition can be divided into 3 types. First, the SE distribution for the small ISIs from 450 to 1,500 ms is focused around a shift in the negative direction with a small spread. This distribution corresponds to anticipatory tapping, i.e., tapping that generates a stable negative asynchrony. As the ISI increased, the dispersion of the distribution increased, and a sharp peak on the positive side occurred in the distribution from 4,800 to 6,000 ms. This positive peak reflects reactive tapping, i.e., tapping that occurs reflexively after hearing the stimulus. Anticipatory tapping with a large negative SE and reactive tapping was mixed in the intermediate ISIs from 1,800 to 3,600 ms. Almost the same distribution was seen under the M condition, but reactive tapping occurred from around 1,800 ms under the M condition with both 4 and 5 words, while reactive tapping occurred with an ISI of 3,600 ms under the N condition.

3.3 Separation of Reactive Tapping and Its Occurrence Rate

Our objective was to obtain information on anticipatory timing control, and we did not analyze reactive tapping that was simply a reflexive movement. For this reason, it was necessary to distinguish between the two types of tapping modes. The examination of the SE distribution for ISI = 6,000 ms shown in Figure 1b demonstrated that almost all the taps were reactive. Since the SE that preceded the auditory stimulus exhibited a large shift in the negative direction, distinguishing between the two types of tapping was relatively simple. Only those taps that were thought to have been reactive in tapping at an ISI of 6,000 ms were selected. The SE mean value of all subjects was calculated based on the SE mean for each subject and was 151 ms below that of the N condition (standard deviation among subjects = 15.7). Thus, the cut-off between the two types of tapping was defined as the value after subtracting 3 times the standard deviation from the mean value. SE = 100 ms was uniformly fixed as the threshold for all subjects and ISIs. SE values larger than this were classified as reactive tapping, and all others were classified as anticipatory tapping.

The percentage of anticipatory tapping observed at each ISI for each subject and the mean among subjects were calculated under the N condition, 4-word condition, and 5-word condition (Fig. 2a). This percentage was defined as the anticipatory-tapping-occurrence rate. Almost 100% of tapping at an ISI below 1,500 ms under the N condition was found to be anticipatory. The anticipatory-tapping-occurrence rate tended to decrease as the ISI increased above 1,800 ms. Mates et al. (1994) found that the time capacity of 2 to 3 s corresponded to the ISI at which reactive tapping begins. It was also found that almost 100% of tapping was anticipatory at an ISI below 1,500 ms under the M condition for both 4 words and 5 words. The anticipatory-tapping-occurrence rate for a higher ISI was smaller than that under the N condition. In addition, if 4- and 5-word conditions are compared, there was almost no difference at a short ISI up to 1,500 ms, but the anticipatory-tapping-occurrence rate at higher ISIs was less for 5 words than for 4 words.

Figure 2b shows the results of a *t* test on the mean value of the anticipatory-tapping-occurrence rate among all subjects for the combinations of N-4 words, N-5 words, and 4–5 words at each ISI. A significant difference in the occurrence rate of anticipatory tapping was observed only at 3,600 ms under the N-4 words condition, whereas a significant difference was observed from 1,800 to 3,600 ms under the N-5 words condition. In addition, the occurrence rate was significantly lower for 5 words at 1,800 compared to that for 4 words. Since the correct response rate under the 5-word condition for the word-memory task was significantly lower than that under the 4-word condition, the N-5 words condition was selected as the dual-task condition to measure the influence of attentional resources.

These findings demonstrate that when tapping is performed with an ISI of 1,500 ms or less, memory tasks are not affected by attentional interference, but are adversely affected with an ISI in the range of 1,800 to 3,600 ms. Furthermore,

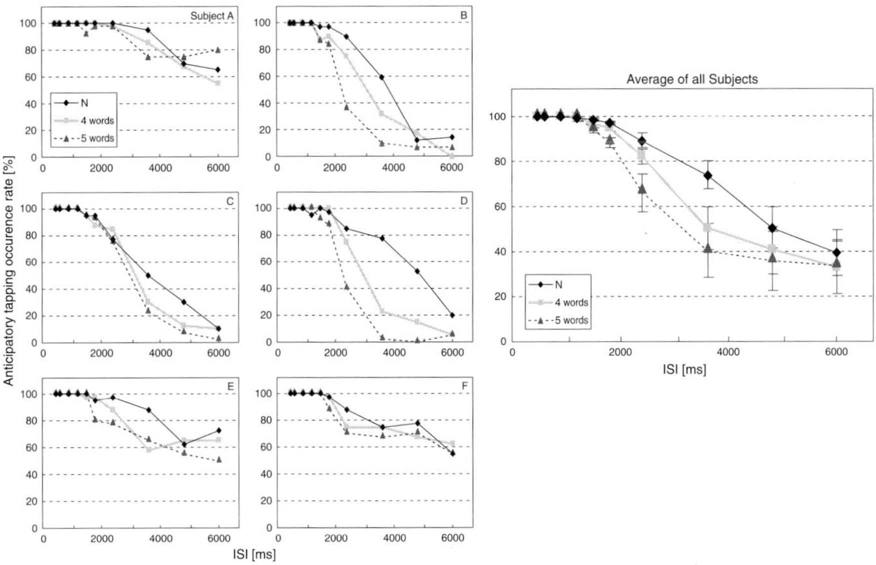

a

t-test of anticipatory tapping occurrence rate

ISI	N–4 words	N–5 words	4–5 words
450	–	–	–
600	–	–	–
900	–		–
1,200			–
1,500		#	
1,800		*	*
2,400	#	*	#
3,000	*	*	
4,800			
6,000			

b

Fig. 2. Occurrence rate of anticipatory tapping. **a** Anticipatory tapping was defined as tapping with an SE less than 100 ms. Left figures show the data from 6 subjects, and the right figure shows the average among 6 subjects. Abbreviations are the same as those in Figure 1. The error bar shows the Standard Error of all subjects. **b** *t*-test of anticipatory tapping occurrence rate. This shows the results of a *t*-test on the mean value of the occurrence rate of anticipatory tapping among all subjects, for the combinations of N-4 words, N-5 words and 4–5 words at each ISI. "*" and "#" indicate significant differences at $P < 0.05$ and $0.05 < P < 0.10$, respectively. The blank column shows other results. We tested all the ISIs except 450, 600, 900 ms (all conditions), and 1,200 ms (4–5 words), because the occurrence rates under these conditions were almost all 100% in this range. (From Miyake et al. 2004, with permission)

with an ISI of 4,800 ms or longer, the effect of attention was small, and the occurrence rate for anticipatory tapping was extremely low. It seems that this region should be considered the domain of reactive tapping, as shown in Figure 1b. It was determined that the synchronization tapping in the stimulus period of 6 s or less can be divided into 3 categories: (i) anticipatory tapping that is unaffected by the subject's attention; (ii) anticipatory tapping that is affected by the subject's attention; (iii) reactive tapping.

However, in the region of 1,800 to 3,600 ms, which is affected by attention, despite an increase in the occurrence rate of reactive tapping under the influence of the memory task (secondary task), not all tapping was reactive. In this ISI range, there was competition between the tapping task and the memory task for the use of attentional resources. This determines the processing efficacy, or, in other words, a "trade-off relationship" exists. This finding corresponds to the "attention capacity hypothesis," which was initially explained.

4 Discussion

The objective of this research was to examine the interference effect of a secondary task on a synchronization tapping task to determine the ISI range that affects attention in the anticipatory timing-control mechanism. The results of this research yielded the following information.

- The negative-asynchrony-occurrence rate was not affected by a secondary task in an ISI range of 450 to 1,500 ms.
- In the ISI range of 1,800 to 3,600 ms, the negative-asynchrony-occurrence rate was significantly reduced by the simultaneous execution of a secondary task.
- The negative-asynchrony-occurrence rate was extremely low in the ISI range of 4,800 to 6,000 ms.

The N condition used in this study was essentially the same as that used in the experiment by Mates et al. (1994). The properties of the SE distribution that are shown in Figure 1b coincide closely with their results. They reported that reactive tapping began to appear at an ISI of 2 to 3 s and that the properties of the negative asynchrony changed in the same range. However, they did not determine the mechanism underlying this phenomenon. The results obtained in the present study using an experiment that took attention into consideration indicated that changes in negative asynchrony depended on two timing mechanisms that qualitatively differ and exist in the ISI regions of 450 to 1,500 ms and 1,800 to 3,600 ms.

The reduction of attentional resources by the execution of a secondary task did not significantly affect the negative-asynchrony-occurrence rate in the 450 to 1,500 ms ISI range. The simultaneous execution of a synchronization tapping task and a secondary task could be within the range of the capacity limit of attentional resources required by both tasks according to the attention-capacity model that was initially proposed. The correct response rate under the 5-word condition for the word-memory task was significantly lower than that under the 4-word condition, where the correct response rate was close to 100% (Table 1). This finding suggests that the attentional resources required to memorize 5 words exceeds or is close to the capacity limit. Therefore, the finding that the tapping task remained unaffected suggests that there is an independent timing-control mechanism for attentional resources in this ISI range.

Movements that can be executed independent of mental processing are referred to as "automatic" (LaBerge and Samuels 1974; Laberge 1975), and regulation of movement through the spinal cord is known to be involved in these movements. For example, there are rhythm generators in the brain stem and spinal cord, such as the central pattern generator (CPG), that produces rhythmic muscle activity like walking (Pearson 1976). These generators are thought to correspond to a timer function that sends periodic pulses in time-perception and production-pacemaker models (Ivry 1996). The possibility has been suggested that tapping in this ISI range is controlled in a feed-forward manner based on the analysis of SE's autocorrelation coefficient (Miyake et al. 2002). It was previously reported that feedback is not received directly from the periphery in the lateral cerebellum, which is responsible for timing control of movement, but that an extremely simple forward control exists (Kawato 1996). These mechanisms may be involved in the automatic anticipatory tapping that was observed in this research.

The synchronization tapping task in the ISI range of 1,800 to 3,600 ms was substantially affected by the lowered attentional resources resulting from the secondary task. However, despite the increase in the occurrence rate of reactive tapping under the influence of the memory tasks, not all tapping became reactive. In addition, a difference was observed in the extent of decrease in the occurrence rate of reactive tapping depending on the number of words to be remembered. These findings indicate a trade-off relationship. The tapping task and the memory task in this ISI range compete with each other for attentional resources and determine the processing efficiency. Consequently, it is necessary to consider what type of processing is involved in the attentional resources that have been diverted by the secondary task to determine the generation mechanism for anticipatory tapping in this ISI range.

The processing that is required in word-memory tasks can be limited to the word-retention activity that accompanies maintenance rehearsal. This type of maintenance rehearsal is thought to be performed by the phonemic loop function, which is a subsystem of working memory (Baddeley 1998a, b). The obtained phonemic information (of a word) is automatically entered in the phonemic storage that is one of the lower-level systems in the phonemic loop and possesses a 1-to-2-s memory buffer. This phonemic storage is related to the maintenance of information concerning rhythm and time intervals (Brown 1997; Saitoh 1997). The phonemic-similarity effect in memory tasks, which is said to be based on the phonemic loop function, has been reported to be lost during the tapping task (Saitoh 1993). The premotor and supplementary motor areas are also involved in the phonemic loop (Osaka 2000), suggesting a relationship between the phonemic loop and motion control.

In this way, the tapping task and word-memory task may compete for the allocation of phonemic storage capacity. This is just a hypothesis, but the fact that stable tapping control is possible in the ISI range of 2 to 3s during a normal tapping task can be explained by this hypothesis. However, if a secondary task results in an overflow in the phonemic storage capacity, time anticipation may become difficult, regardless of the ISI. The results of this research, in which there was no apparent influence of the memory task at ISIs of 1,500ms or less, contradicts this hypothesis. We propose that anticipatory timing control is achieved through the interaction between time perception based on phonemic storage and automatic movement mechanisms in the actual timing control.

Our research was aimed at furthering psychological analyses related to the time-perception mechanism in anticipatory timing synchronization, which is thought to be indispensable in cooperative activity among humans. The results demonstrated for the first time the presence of two types of anticipatory mechanisms in the synchronization tapping task from the standpoint of attention involved in time perception. One is anticipatory tapping influenced by attention and seen at the ISI range of 1,800 to 3,600ms, and the other is the automatic tapping mechanism that is not affected by attention and is seen at the 450-to-1,500-ms range. Accordingly, this anticipatory timing mechanism can be considered a dual process in which the anticipatory mechanisms work together based on the processing of the implicit automatic anticipation and the explicit processing of temporal information.

Finally, exactly how this type of perception- and movement-integrative process is involved in higher-level brain functions, such as attention and awareness, is an extremely complex problem. Pöppel et al. have already tackled the problem of integrating information in the temporal region through the framework of a "time window" (Pöppel 1971, 1988, 1997; Szelag et al. 2002). Humans integrate information in this 3-s time window and generate a state of awareness that corresponds to a "subjective present." The anticipatory timing mechanism is closely related to this type of temporal integration, and the findings of this study suggest that this time window is formed by a dual process of anticipation. If the physiologic foundation for this temporal-perception mechanism can be clarified through imaging techniques such as f-MRI, it may be possible to construct a model for the neuronal mechanism demonstrated in this study. We also expect this to be related to the technology that supports cooperative processes among humans within the range of cognitive time.

References

Aschersleben G, Prinz W (1995) Synchronizing actions with events: the role of sensory information. Percept Psychophys 57:305–317

Baddeley A (1986) Working memory. Oxford University Press, New York

Baddeley A (1998a) Working memory. Comptes Rendus de l'Academie des Sciences – Series III – Science de la Vie 321:167–173

Baddeley A (1998b) Recent developments in working memory. Curr Opin Neurobiol 8:234–238

Brown SW (1997) Attentional resources in timing: interference effects in concurrent temporal and nontemporal working memory tasks. Percept Psychophys 59:1118–1140
Daneman M, Carpenter PA (1980) Individual differences in working memory and reading. J Verb Learn Verb Behav 19:450–466
Fraisse P (1966) The sensorimotor synchronization of rhythms. In: Requin J (Eds) Anticipation et comportement. Centre National, Paris, pp 233–257
Ivry RB (1996) The representation of temporal information in perception and motor control. Curr Opin Neurobiol 6:851–853
Ivry RB (1997) Neural mechanisms of timing. Trends Cogn Sci 1:163–169
Kagerer FA, Wittmann M, Szelag E, Steinbüchel N (2002) Cortical involvement in temporal reproduction: evidence for differential roles of the hemispheres. Neuropsychologia 40:357–366
Kahnemann D (1973) Attention and efforts. Prentice-Hall, Englewood Cliffs NJ
Kawato M (1996) Computational Theory of Brain (In Japanese). Sangyo Tosho Publisher, Tokyo
Kolers PA, Brewster JM (1985) Rhythms and responses. J Exp Psychol Hum Percept Perform 11:150–167
LaBerge D, Samuels SJ (1974) Toward a theory of automatic information processing in reading. Cognit Psychol 6:293–323
Laberge D (1975) Acquisition of automatic processing of perceptual and associative learning. In: Rabbitt PMA, Dornic S (Eds) Attention and performance V. Academic Press, New York
Macar R, Casini L (1999) Multiple approaches to investigate the existence of an internal clock using attentional resources. Behav Process 45:73–85
Mangles JA, Ivry RB, Shimizu N (1998) Dissociable contributions of the prefrontal and neocerebellar cortex to time perception. Cogn Brain Res 7:15–39
Mates J, Radil T, Müler U, Pöppel E (1994) Temporal integration in sensorimotor synchronization. J Cogn Neurosci 6:332–340
Miyake Y, Heiss J, Pöppel E (2001) Dual-anticipation in sensory-motor synchronization. Proceedings of 1st Int. Symp. on Measurement, Analysis and Modeling of Human Functions (ISHF2001), Sapporo, Japan, pp 61–66
Miyake Y, Onishi Y, Pöppel E (2002) Two modes of timing anticipation in synchronization tapping (in Japanese). Transaction of SICE 38:1114–1122
Miyake Y, Onishi Y, Pöppel E (2004) Two types of anticipation in synchronous tapping. Acta Neurobiol Exp 64:415–426
Osaka N (2000) Brain and working memory. Kyoto University Press, Koyoto
Osaka M, Osaka N (1994) Working memory capacity related to reading: measurement with the Japanese version of reading span test. Jpn J Psychol 65:339–345
Pascual-Leone A (2001) Increased variability of paced finger tapping accuracy following repetitive magnetic stimulation of the cerebellum in humans. Neurosci Lett 306:29–32
Pearson K (1976) The control of walking. Sci Am 235:72–86
Peters M (1989) The relationship between variability of intertap intervals and interval duration. Psychol Res 51:38–42
Pöppel E (1971) Oscillation as possible basis for time perception. Studium Generale 24:85–107
Pöppel E (1988) Mind works: time and conscious experience. Harcourt Brace Jovanovich, Boston MA
Pöppel E (1997) A hierarchical model of temporal perception. Trends Cogn Sci 1: 56–61

Rao SM, Harrington DL, Haaland KY, Bobholz JA, Cox RW, Binder JR (1997) Distributed neural systems underlying the timing of movements. J Neurosci 17:5528–5535
Saitoh S (1993) The disappearance of the phonological similarity effect by complex rhythmic tapping. Psychologia 36:27–33
Saitoh S (1997) Research of phonetic working memory (in Japanese). Fuhma Shobo Publisher, Tokyo
Stevens LT (1886) On the time sense. Mind 11:393–404
Szelag E, Kowalska J, Rymarczyk K, Pöppel E (2002) Duration processing in children as determined by time reproduction: implications for a few seconds temporal window. Acta Psychol 110:1–19
Woodrow H (1932) The effect of rate of sequence upon the accuracy of synchronization. J Exp Psychol 15:357–379
Zelaznik HN, Spencer RMC, Ivry RV (2002) Dissociation of explicit and implicit timing in repetitive tapping and drawing movements. J Exp Psychol Hum Percept Perform 28:575–588

Subject Index